GCI Scientific Program Reports

Seismic Stabilization of Historic Adobe Structures

Final Report of the Getty Seismic Adobe Project

E. Leroy Tolles

Edna E. Kimbro

Frederick A. Webster

William S. Ginell

The Getty Conservation Institute Los Angeles

Seismic Stabilization of Historic Adobe Structures

Final Report of the Getty Seismic Adobe Project

Dinah Berland, *Editorial project manager*
Dianne Woo, *Manuscript editor*
Anita Keys, *Production coordinator*
Garland Kirkpatrick, *Cover designer*
Hespenheide Design, *Designer*

Illustrations of models in chapter 7 courtesy of IZIIS, University "SS. Cyril and Methodius," Skopje, Republic of Macedonia.

Printed in the United States of America
10 9 8 7 6 5 4 3 2 1

The Getty Conservation Institute works internationally to advance conservation in the visual arts. The Institute serves the conservation community through scientific research, education and training, model field projects, and the dissemination of information. The Institute is a program of the J. Paul Getty Trust, an international cultural and philanthropic institution devoted to the visual arts and the humanities.

The GCI Scientific Program Reports series presents current research being conducted under the auspices of the Getty Conservation Institute.

Library of Congress Cataloging-in-Publication Data

Seismic stabilization of historic adobe structures : final report of the Getty Seismic Adobe Project / E. Leroy Tolles . . . [et al.].

p. cm.—(GCI scientific program reports)

Includes bibliographical references.

ISBN 0-89236-587-0

1. Building, Adobe. 2. Buildings—Earthquake effects. 3. Earthquake engineering. 4. Historic Buildings—Southwest (U.S.)—Protection. I. Tolles, E. Leroy, 1954– II. Getty Adobe Project. III. Getty Conservation Institute. IV. Series.

TH4818.A3 S45 2000

693′.22—dc21 00-022015

CIP

Contents

ix **Foreword** *Timothy P. Whalen*
xi **Preface** *William S. Ginell*
xv **Acknowledgments**
xvii **Project Participants**

Chapter 1
1 **Background to the Getty Seismic Adobe Project**
1 Getty Seismic Adobe Project Goals and Purpose
2 Life-Safety Issues for the Seismic Retrofit of Adobe Buildings
2 Conservation Issues for the Seismic Retrofit of Adobe Buildings
4 Seismic Performance and Seismic Retrofit
8 Project Approach
First-year activities
Second-year activities
Third-year activities

Chapter 2
11 **Overview and Procedures for Models 1–9**
11 Seismic Retrofit Techniques
13 Overview of Tests
14 Description of Materials and Models
Adobe material
Model design and construction
Retrofit measures
Model similitude
17 Description of Test Procedure
Test setup
Instrumentation
Simulated earthquake motions
Documentation

Chapter 3
21 **Description of Tests for Models 4–6**
21 Description of Models
Model 4 (S_L = 5)
Model 5 (S_L = 11)
Model 6 (S_L = 11)
23 Test Results for Model 4
28 Test Results for Model 5
31 Test Results for Model 6
37 Summary of Test Results for Models 4–6

Chapter 4 | 41 | **Description of Tests for Model 7**
41 Layout of Model 7
42 Retrofit Measures
45 Summary of Test Results for Model 7

Chapter 5 | 53 | **Description of Tests for Models 8 and 9**
53 Layout of Models 8 and 9
55 Model 8 Retrofit Measures
58 Test Results for Model 8
58 Flexural Test of a Wall Element with Center-Core Reinforcement
59 Test Results for Model 9
70 Summary of Test Results for Models 8 and 9

Chapter 6 | 73 | **Analysis of Test Results for Models 1–9**
73 Analysis of Model Test Results
Initial crack development (test levels II–IV)
Performance during moderate to strong seismic levels (test levels V–VII)
Performance during very strong seismic levels (test levels VIII–X)
Summary of performance of model buildings
86 Summary of Field Observations after the Northridge Earthquake
87 Comparison of Laboratory Results and Field Observations

Chapter 7 | 89 | **Description of Tests for Models 10 and 11**
91 Overview of Tests
92 Materials Tests
92 Dynamic Testing Procedures
93 Model Buildings

Chapter 8 | 103 | **Test Results and Data Analysis for Models 10 and 11**
104 Model 10—Unretrofitted
Elastic response
Damage progression
Performance and collapse during severe ground motions
111 Model 11—Retrofitted
Elastic response
Damage progression
Performance during severe ground motions
119 Loads on Elements of the Retrofit System
Analysis and discussion of results
Comparison of elastic response
Comparison of damage progression
Comparison of performance during severe ground motions
134 Performance Observations and Summaries
138 Summary of Comparison of Performance of Small- and Large-Scale Models

Chapter 9 | 141 | **Recent Application of GSAP Technology**

Chapter 10 145 **Summary and Conclusions**
146 Assessment of Retrofit Measures
147 Relative Model Performance

Chapter 11 149 **Suggestions for Future Research**

151 **References**

153 **Glossary**

155 **Bibliography of Publications Resulting from GSAP**

157 **About the Authors**

Foreword

There is something satisfying and appealing about the use of earth as a building material. An earthen building draws its raw material from the ground on which it stands and eventually returns to the earth—the ultimate recyclable and renewable resource. Probably the first of humankind's building materials, earth has been in continuous use in various guises throughout history and on every continent. Of the many distinct ways in which earth has been used to construct human habitations, adobe may be the simplest. The word *adobe* comes from the Arabic word for "brick" and was introduced to the Americas, along with adobe technology itself, by the Spanish. Mud brick was widely adopted in the Southwest as a building material: the simplicity of the technology and perhaps absence of timber or kilns as well as the region's bright sunshine and dry desert air were highly conducive to its use.

As is commonly known, however, adobe structures are highly vulnerable to earthquakes. Categorized by architects and builders as unreinforced masonry, and the weakest type of structure in that category, adobe buildings have been devastated in areas of high seismicity. As pointed out in this book, of the approximately 900 adobes originally constructed in the San Francisco Bay Area alone, the number had dwindled to about 65 by the 1940s. Today only some 350 historic adobes remain in California, many of which date to the Spanish colonial period.

The Getty Conservation Institute has, since its inception, addressed the threats of seismic destruction of cultural property. This is not simply because Los Angeles itself is in a seismic zone but rather because many seismic areas in the world—the west coast of North and South America, areas of the Mediterranean, and China, to name but a few—contain enormous concentrations of cultural patrimony. Among the first books published by the GCI was *Between Two Earthquakes: Cultural Properties in Seismic Zones* by Sir Bernard Feilden (1987). The GCI has conducted research on securing museum objects, ranging from extremely simple tie-down devices to highly sophisticated engineering mounts, and has sponsored seminars and conferences, as in Quito and Cairo, on earthquake response. The present book represents a further step in our commitment to the issue and is the fruit of a major research project of the GCI.

The Getty Seismic Adobe Project (GSAP) initiative was first discussed at the Sixth International Conference on Earthen Architecture,

"Adobe 90," held in October 1990 in Las Cruces, New Mexico, where the GCI and the Museum of New Mexico State Monuments had been conducting long-term research and field testing at the historic monument of Fort Selden, an adobe frontier fort dating from the mid-nineteenth century. The 1989 Loma Prieta earthquake in California, which had severely damaged or destroyed historic adobe buildings in the area, was fresh in the minds of delegates, as it emphasized once again how each significant earthquake destroys or degrades the authenticity of historic structures.

Because the earthquake codes for seismic stability in California and elsewhere require invasive retrofitting to stabilize historic structures—which in the case of adobe results in a severe mismatch of fabric—the primary objective of the GSAP initiative was to find technologically sympathetic and minimally invasive methods of stabilizing these structures. The objective was not to make adobes "earthquake-proof" but rather to ensure safety by preventing the overturning of walls during a seismic event. As a result of ten years of research conducted with the help of an international group of specialists and earthquake engineers, this objective has been met. The research was conducted first on small-scale models at Stanford University and later on larger-scale models in Skopje in the Republic of Macedonia and truly represents an exemplar of collaborative enterprise. The first applications of GSAP research are now taking place in California, reflecting the ultimate purpose of conservation research as a whole: to identify the problem, research its solutions, and disseminate the results—thus ensuring, insofar as possible, the application of tested methods in the field.

It is particularly gratifying for me to present this important publication, which is applicable not only to preservationists, engineers, and architects but also to a wider community of private owners of earthen buildings both in the Americas and elsewhere in the world.

Timothy P. Whalen
Director, The Getty Conservation Institute

Preface

Many of the early structures that date back to the Spanish colonial period in the southwestern United States were built of mud brick, or adobe. The materials for construction of the early churches, missions, and houses were generally limited to those that were available in the region and easily worked by local artisans. Adobe has many favorable characteristics for construction in arid regions: it provides effective thermal insulation, the clayey soil from which adobe bricks are made is ubiquitous, the skill and experience required for building adobe structures is minimal, and building construction does not require the use of scarce fuel. As a consequence of their age, design, and functions, surviving adobe buildings are among the most historically and culturally significant structures in their communities.

The seismic vulnerability of California's adobe architectural heritage—its missions, presidios, and residential structures—has been recorded in mission documents since the eighteenth century. Some of the existing adobe mission churches have been reconstructed many times on the same site following the destruction of the preceding structure by an earthquake. For example, the 1971 San Fernando earthquake in Los Angeles County destroyed the San Fernando Mission church and severely damaged its elegant *convento* (priest's residence), the largest and most elaborately decorated residential adobe structure ever erected in California. The Native American neophyte mural paintings of the convento were lost forever.

In recent years, considerable progress has been made toward understanding the behavior of buildings during earthquakes. Unreinforced masonry buildings are widely recognized as being particularly vulnerable. California law now mandates that within this decade, local jurisdictions reduce the hazard posed by existing buildings, especially those constructed of unreinforced masonry. Adobe structures fall within the class of unreinforced masonry buildings and are particularly susceptible to earthquake damage. Many unreinforced masonry buildings either are being retrofitted to improve stability and to decrease the possibility of loss of life, or are being demolished.

The retrofitting of the vast majority of commercial and residential buildings poses only economic and technical problems, and conventional engineering approaches are well suited to such buildings. When planning reduction of the seismic vulnerability of historic and culturally significant buildings, however, ethical considerations that limit the design alternatives must be made. Engineering interventions to improve a structure's seismic safety can lead to irreparable loss or damage to the historic fabric of the building and to its architectural or decorative features.

Many previously damaged, culturally significant, and vulnerable adobe structures are in need of structural modification that will improve their likelihood of survival. The difficulties in implementing these modifications lie in the often conflicting requirements of using conventional engineering practices while maintaining, intact, the historical and cultural features.

When confronted by the necessity of action in face of the seismic threat, authorities responsible for the California missions and other historic adobe buildings and earthen architectural monuments are forced to choose from among four courses of action for these structures:

1. closing and fencing off the buildings, thereby beginning the inevitable progressive deterioration caused by lack of use and maintenance;
2. demolishing the buildings, as has been done to many historic structures that fail to comply with earthquake safety ordinances;
3. retrofitting them using the presently available, highly destructive, invasive, and expensive measures sanctioned for adobe structures; or
4. retrofitting them using innovative, tested techniques specifically developed for adobe structures and designed to observe the conservation principle of minimal intervention.

The fourth course of action is the only alternative that will preserve the cultural and historic value of adobes while providing life safety to the occupants of the buildings.

In keeping with the ongoing commitment of the Getty Conservation Institute to the preservation of our collective cultural heritage, the Getty Seismic Adobe Project (GSAP) was initiated in November 1990. The goal of the project was to develop technical procedures for improving the seismic performance of historic adobe structures consistent with providing life safety and maintaining architectural, historic, and cultural conservation values.

Overview of This Book

This book is not only the final report of GSAP activities, it is also the first publication to provide an overview of the results of scale-model laboratory research along with field data from a survey of damage to historic adobe buildings after an actual earthquake.

The principal part of this book contains a summary of the small- and large-scale shaking-table tests and an analysis of results.

These results are considered in relation to field observations of seismic damage to adobes, particularly as detailed in *Survey of Damage to Historic Adobe Buildings after the January 1994 Northridge Earthquake* (Tolles et al. 1996). The text summarizes the results of preliminary studies that were described in internal project reports (Thiel et al. 1991; Tolles et al. 1993) and also provides details of recent experimental work that validated an innovative approach to the problem of the seismic retrofitting of adobe structures.

Chapter 1 provides an overview of the Getty Seismic Adobe Project, its research goals, project approach, and activities. Chapter 2 summarizes the small-scale-model (1:5) testing program and procedures used at the shaking table at the John A. Blume Earthquake Engineering Center at Stanford University in Palo Alto, California. The results of the tests on models 4–6 are presented in chapter 3; the results for model 7 are given in chapter 4; and the data on models 8 and 9 are presented in chapter 5. Chapter 6 contains an analysis and a synthesis of the significant aspects of the behavior of models 1–9.

In chapter 7, the need for large-scale-model (1:2) tests is discussed, along with the procedures and results obtained on the large-scale models 10 and 11. These tests were performed at the Institute of Earthquake Engineering and Engineering Seismology (IZIIS), University "SS. Cyril and Methodius," in Skopje, Republic of Macedonia. An analysis of the results of the large-scale tests is given in chapter 8. Chapter 9 briefly describes one recent application of GSAP technology to the retrofitting of adobe structures in California. The conclusions reached in this study and suggestions for future work are provided in chapter 10. At the end of the book are a glossary of terms related to adobe structures and seismic retrofitting, and a selected bibliography of journal articles, reports, and other publications generated as a result of GSAP research.

A forthcoming volume, *Planning and Engineering Guidelines for the Seismic Stabilization of Historic Adobe Structures* by Tolles, Kimbro, and Ginell, will provide information on how to plan for and access further information on the retrofitting of historic adobe buildings.

William S. Ginell

Acknowledgments

The authors would like to thank the many colleagues who contributed to the successful completion of this research program. First, they would like to thank Charles C. Thiel Jr. for his sound professional guidance during the first few years of this research program. His contributions were critical to the design and development of the concepts of the program and the implementation of the small-scale model tests.

They also wish to thank Helmut Krawinkler and Anne Kiremidjian, who, during the course of this project, held positions as directors at the John A. Blume Earthquake Engineering Center, Stanford University. Their support and contributions are greatly appreciated. Krawinkler was also a member of the GSAP Advisory Committee, and his contributions in that capacity are also very much appreciated.

Also at Stanford University, Chris Thomas was responsible for the construction of the nine small-scale model buildings that were tested during the research program. His hard work and attention to detail were extremely valuable to those of us who designed the tests that destroyed his beautiful models.

The large-scale models were built and tested under the supervision of Predrag Gavrilovic at IZIIS, University "SS. Cyril and Methodius," in Skopje, Republic of Macedonia. He was assisted by Veronika Sendova, Ljubomir Taskov, Lidija Krstevska, and the competent staff of engineers and technicians at IZIIS. The authors are indebted to all of them for their contributions to the success of the final stages of GSAP.

The members of the GSAP Advisory Committee (listed on page xviii) also played a very important role in the research process. Their valuable contributions to the project and its research goals have helped the results to become more applicable to ongoing retrofit efforts.

The authors are particularly grateful for the support at the Getty Conservation Institute of Neville Agnew, principal project specialist and the initial project director for GSAP; and Frank Preusser, former head of the GCI Scientific Program, for their strong and continued support throughout this entire research project. Their interest and encouragement were critical to the success of GSAP.

Thanks to Getty Conservation Institute staff members Ford Monell, whose excellent work in assembling the original manuscript is much appreciated; Jacki Redeker, who organized the many electronic images; and Valerie Greathouse, who provided bibliographic reference assistance. Appreciation is also extended to the staff of Getty Trust Publication Services who helped bring this book to light, particularly Dinah Berland, editorial project manager for the book, and Anita Keys, production coordinator; and to consultants Dianne Woo, who edited the manuscript, and Scott Patrick Wagner, who edited the references and prepared the manuscript for production.

Project Participants

The following individuals actively participated in the Getty Seismic Adobe Project during various periods from 1990 through 1999.

Neville Agnew	Principal project specialist Getty Conservation Institute, Los Angeles
Anthony Crosby	Historical architect Formerly with the National Park Service, Denver
Predrag Gavrilovic	Professor, civil engineering IZIIS, University "SS. Cyril and Methodius" Skopje, Republic of Macedonia
William S. Ginell	Senior scientist Getty Conservation Institute, Los Angeles
Edna E. Kimbro	Architectural conservator and historian Watsonville, Calif.
Helmut Krawinkler	Professor, structural engineering Stanford University, Palo Alto, Calif.
Lidija Krstevska	Civil engineer Institute of Earthquake Engineering and Engineering Seismology (IZIIS), University "SS. Cyril and Methodius" Skopje, Republic of Macedonia
Veronika Sendova	Assistant professor, civil engineering IZIIS, University "SS. Cyril and Methodius" Skopje, Republic of Macedonia
Ljubomir Taskov	Associate professor, civil engineering IZIIS, University "SS. Cyril and Methodius" Skopje, Republic of Macedonia
Charles C. Thiel Jr.	Seismic structural engineer Piedmont, Calif.
Chris Thomas	Engineering laboratory assistant Stanford University, Palo Alto, Calif.

E. Leroy Tolles	Seismic structural engineer Earthen Buildings Technologies Lafayette, Calif.
Frederick A. Webster	Civil/structural engineer Menlo Park, Calif.

GSAP Advisory Committee

Edward Crocker	New Mexico Community Foundation Santa Fe, N.Mex.
Anthony Crosby	Historical architect Formerly with the National Park Service, Denver
M. Wayne Donaldson	Historical architect San Diego, Calif.
Melvyn Green	Structural engineer Manhattan Beach, Calif.
James Jackson	Architect California Department of Parks and Recreation Sacramento, Calif.
Helmut Krawinkler	Professor, structural engineering Stanford University, Palo Alto, Calif.
John Loomis	Architect Newport Beach, Calif.
Nicholas Magalousis	Former curator, Mission San Juan Capistrano Laguna Beach, Calif.
Julio Vargas Neumann	Professor, structural engineering Pontifica Universidad Católica del Peru, Lima

Chapter 1
Background to the Getty Seismic Adobe Project

Getty Seismic Adobe Project Goals and Purpose

The Getty Seismic Adobe Project (GSAP) was initiated by the Getty Conservation Institute in November 1990 for the purpose of developing technical procedures for improving the seismic performance of historic adobe structures with minimal impact on the historic fabric of these buildings. The program focused on the Spanish colonial missions and historic adobes in seismic areas of the southwestern region of the United States, with expected applications to historic adobes in other seismic regions, particularly Central and South America.

Many historic adobe buildings have fared very poorly during earthquakes. The seismic behavior of adobe buildings—as well as those made of stone and other types of unreinforced masonry—is commonly characterized by sudden and dramatic failure. A high likelihood of serious injuries and loss of life usually accompanies the local or general collapse of such structures. Generally, the evaluation of the engineering community is that adobe buildings, as a class, pose the highest risk among the various building types. Nevertheless, it has been observed that some adobes have withstood repeated, severe earthquake ground motions without total collapse.

The seismic upgrading of historic buildings embraces two distinct goals: (1) seismic retrofitting to provide adequate life-safety protection, and (2) preserving the historic (architectural) fabric of the building. These goals are often perceived as fundamentally opposed. In current seismic retrofitting practices, substantial alterations of structures are usually required, involving new structural systems and often substantial removal and replacement of building materials. Historic structures that are strengthened and fundamentally altered in this manner lose much of their authenticity. They are often virtually destroyed because of their presumed earthquake risk without, or instead of, waiting for an earthquake to damage them. As a result, the conflict is seen as an either/or proposition: either the building can be retrofitted, making it safe during seismic events but destroying much of the historic fabric in the retrofitting process, or the building with its historic fabric can be left intact while the risk of potential structural failure and collapse during future seismic events is accepted.

Faced with the apparent conflict between the unacceptable seismic hazard posed by many adobe buildings and the unacceptable conservation consequences of expensive, conventional retrofitting approaches, the Getty Conservation Institute made a serious commitment to research and develop seismic retrofitting approaches for historic adobe structures that would balance the need for public safety and the conservation of these cultural assets. The long-term goal of the project was to develop and evaluate design practices and tools that could be made available to architects, engineers, owners, and building officials. From the outset, it was understood that the project's success would be measured in terms of the extent of the application of GSAP technologies and their effectiveness in achieving the dual goals of seismic safety and maintenance of historic fabric.

Life-Safety Issues for the Seismic Retrofit of Adobe Buildings

A fundamental goal of building regulations is to provide for adequate life safety during the most severe seismic events. A building that is a total economic loss in an earthquake still may be judged a success if the damage to the structure poses little life-loss hazard to its occupants. The intention of modern building codes is to prevent structural damage during moderate earthquakes, but structural damage still may occur during seismic events of greater magnitude. Except for the most important facilities, buildings are designed with the assumption that major earthquakes will cause some structural damage.

The first objective of the seismic retrofit measures developed as part of GSAP was to minimize the life-loss hazard. Structural damage may occur, and cracks in the walls may develop, but it is essential to provide for public safety by preventing structural instability and other damage that may cause injury or loss of life. Seismic retrofit measures that minimized the risk of life loss and also satisfied basic conservation criteria—minimal intervention and reversibility—were judged as successful.

Once these measures were identified, the next important objective of the project was to minimize other types of damage. Some seismic retrofit measures may provide for life safety but have little effect on preventing cracking during moderate earthquakes and may allow significant and nonreparable damage during major seismic events. Other measures may reduce cracking during moderate events but have a negligible effect on life-threatening instability during major events.

Conservation Issues for the Seismic Retrofit of Adobe Buildings

As earthquakes continue to occur in California at greater magnitudes and with more damaging effects than those of the last seventy-five years, the substantial seismic hazard posed by historic adobe structures is likely to become more widely known and understood. Public officials are unlikely to allow the continued use of buildings clearly at high risk, regardless of their historical importance. There appear to be only three options to preserving these historic structures:

1. severely restrict the use of historic adobe buildings, allowing them to be observed only from a safe distance;
2. seismically strengthen the buildings using current, strength-based earthquake engineering practices, which can substantially alter their historic fabric and reduce their authenticity; or
3. develop new approaches to the design of retrofits that are specifically adapted to the nature of adobe and its use in historic buildings, and that have minimal and reversible impact on the historic fabric of the buildings.

Historic preservation involves more than just maintaining the buildings as artifacts on display. The opportunity to use and experience a historic building as it was originally intended is integral to its preservation. GSAP was undertaken by the Getty Conservation Institute to develop the third option so that seismic retrofitting does not compromise a building's cultural significance.

The conservation approach for adobe structures described here involves a focused, disciplined development of design options consistent with preserving the building's historic fabric. This is a four-step process: first, the structure is fully characterized; second, important features and significant characteristics are identified; third, an understanding of the structure in context is developed; only then can the fourth step be undertaken of developing design options that are respectful of the structure's historic fabric.

Once life safety is established, the issue of limiting the extent of damage to adobe buildings during seismic events is then addressed. The preservation of historic adobes is important not only before the next major earthquake but also afterward. First, damage must be limited to reparable levels during the most severe earthquakes. The next step is to limit the amount of cosmetic damage during moderate earthquakes. The objectives of seismic retrofit measures that satisfy conservation criteria are ranked in order of importance as follows:

1. Provisions for life safety during the most severe earthquakes
2. Limitation of damage to reparable levels during the most severe earthquakes
3. Minimizing damage during moderate earthquakes

Different retrofit measures may be used to satisfy each of these objectives. The life-safety objective must be ranked first, but the second and third objectives are interchangeable depending on the goals of the decision makers. For example, it may be more important to prevent cosmetic damage to surface finishes during frequently occurring moderate earthquakes than to ensure that a building remains reparable during infrequent, major temblors.

Seismic Performance and Seismic Retrofit

In countries where a large percentage of the houses are constructed of adobe, the number of deaths following a major earthquake often reaches thousands or tens of thousands. Despite their poor performance during earthquakes, many adobes have withstood repeated, severe ground motions without catastrophic collapse. The Castro Adobe in Watsonville, California, for example, is a two-story rectangular building with thick walls and two interior cross walls. It has survived the 1865 earthquake in the Santa Cruz Mountains, the 1906 San Francisco earthquake, and the 1989 Loma Prieta earthquake in Santa Clara County. Although damage was significant in 1989—with reactivation of cracks in the long walls, movement of cracked sections, and collapse of the upper portion of one gable-end wall—the principal elements of the structure remained standing, stable, and reparable. Simple retrofit measures could have prevented the gable wall failure. A similar history can be related for Mission Dolores in San Francisco and the Serra Chapel in San Juan Capistrano, California, among others. Adobe structures can exhibit acceptable behavior if they are well maintained (kept dry) and have the right structural characteristics.

The January 17, 1994, Northridge earthquake in the Los Angeles metropolitan area was a true test of the seismic performance of historic adobe buildings. This earthquake caused more damage to historic adobe buildings than any other since the 1925 Santa Barbara earthquake. Three important Southern California buildings suffered serious generalized damage, including at least one wall collapse. They were the De la Osa Adobe, Encino; the Andres Pico Adobe, Mission Hills; and the Del Valle Adobe at Rancho Camulos, near Piru. Several other buildings suffered considerable damage. The details of these and other damaged adobes are covered in *Survey of Damage to Historic Adobe Buildings after the January 1994 Northridge Earthquake* (Tolles et al. 1996).

Understanding the seismic performance of structures in terms of engineering science is of recent vintage. Only in the twentieth century did information begin to emerge on how structures respond in earthquakes. Historical building practices developed with the accumulation of experience gained through trial and error. The first measurements of ground motions in damaging earthquakes were not taken until 1933, and it was not until the 1970s that the first recordings were made of a building as it responded to an earthquake that caused damage to that structure. The first procedures for seismic design were not formulated until early in the twentieth century, although there had been some sporadic attempts prior to that time. Many assorted construction details were proposed that were asserted to provide better seismic performance. Following the emergence of modern construction methods in which steel and reinforced concrete replaced brick and stone as principal building materials, structural designs were developed that could withstand environmental loads (wind and earthquake) and perform in a relatively predictable and acceptable manner. Steel and reinforced concrete are ductile materials that have linear elastic properties and good post-elastic strength characteristics. After yielding, these materials maintain most of their strength while undergoing substantial plastic deformations. They can be analyzed with reasonable accuracy using analytical or computational methods. In contrast, the behavior of brittle, unreinforced materials—such as stone, brick, or adobe—is extremely difficult to predict after cracks are initiated, even with today's advanced computational capabilities. Even if results could be generated with these technologies, they would not be accurate. Once yielding occurs in a brittle material, cracks develop, and a complete loss of tensile strength results. The seismic behavior of adobe buildings after cracks have developed is dominated by the interactions of large, cracked sections of walls that rock out of plane and collide against each other in plane.

A conceptual revolution in seismic design occurred in the 1960s. Engineers began to develop the notions of ductile design—that is, the ability of a structural system to continue to support gravity and to reverse seismic loads after the building materials have yielded. Prior to the development of this notion, the essential approach to seismic design was to provide strength to resist the lateral loads in the structure. Ductile design approaches have not abandoned strength concepts; instead, they have been supplemented by implementing reinforcement and connection detailing so that elements have the capacity to transmit loads even after they have been damaged. In its simplest form, the term *ductility* has come to mean the ratio of the displacement at which failure occurs to that at which yielding occurs (permanent deformation). Steel is characterized as a highly ductile material, as is reinforced concrete when the reinforcing is properly placed. Brittle materials (e.g., fired brick, adobe, tile, glass, and unreinforced concrete), while they may have large compressive strengths, characteristically have low ductility unless reinforced. Unreinforced adobe has low material ductility coupled with low strength, which is generally stated as the reason for its poor seismic performance.

The standard criteria for typical seismic design are (1) design the structure to remain elastic during moderate to major seismic events, and (2) design the individual elements and connections of the structure to perform in a ductile manner and retain their strength during major seismic events (Wiegel 1970). In such an approach, the design of the structure in the post-elastic phase is not explicitly analyzed; criteria for the design of concrete and steel construction are based on a combination of field experience and laboratory experimentation.

Because adobe is a brittle material, the fundamentals of its post-elastic behavior are entirely different from those of ductile building materials. Once a typical unreinforced adobe wall has cracked and the tensile strength of the wall is lost, the wall can continue to carry vertical loads as long as it remains upright and stable. Cracks in adobe walls may occur from seismic forces, from settlement of the foundation, or from internal loads (e.g., roof beams). Even though the tensile strength of the wall material has been lost, the structure still may remain standing. The thickness of typical historic adobe walls makes these walls difficult to destabilize even when they are severely cracked. The support provided at the tops of the walls by a roof system may add stability, especially when the roof system is anchored to the walls. In many adobe buildings, the height-to-thickness (slenderness) ratios (S_L) may be less than 5, and the walls can be 1.2–1.5 m (4–5 ft.) thick, both of which make wall overturning unlikely. Seismic retrofit techniques can be used to improve structural stability and reduce the differential displacements of the cracked sections of the structure.

Previous dynamic laboratory research on the seismic behavior of adobe structures has been performed only in few areas of the world despite the global nature of the problem. The first shaking-table tests were performed in Mexico during the 1970s (Meli, Hernandez, and Padilla 1980). Five model adobe buildings, 1:2.5 in scale, were tested. The buildings were modified to include a concrete bond beam, horizontal steel rods, and welded wire mesh applied to the exterior adobe surface.

Extensive research on the strengthening of adobe construction has been performed at the Catholic University in Lima, Peru. The work was focused largely on the determination of materials properties but has also included tilt-table tests (Vargas-Neumann and Otazzi 1981), shaking-table tests (Vargas-Neumann, Bariola, and Blondet 1984), and dynamic tests on the out-of-plane stability of adobe walls (Bariola Bernales 1986). The emphasis of the shaking-table tests was on the seismic stabilization of new adobe construction whose walls were reinforced with internal canes and wood bond beams.

Dynamic tests were conducted in the United States during the 1980s supported by grants from the National Science Foundation. Six roofless adobe model structures, 1:5 in scale, were tested at Stanford University, Palo Alto, California, to evaluate the effects of simple retrofit techniques on their dynamic behavior up to and including collapse (Tolles and Krawinkler 1989). Large-scale model tests were conducted at the University of California, Berkeley, on adobe models fitted with wood bond beams and various types of wire mesh attached to wall surfaces (Scawthorn and Becker 1986).

Many seismic retrofits of adobe buildings attempt to strengthen the material through application of reinforcing products or addition of ductile, reinforced elements that allow the structural elements to maintain strength during severe seismic activity. One example is replacement of the center of an adobe wall with reinforced concrete (e.g., Sonoma Barracks, Sonoma State Historic Park, Sonoma County, Calif.). Such a design is based on the requirement that the wall elements must retain strength and ductility, which is a standard elastic design criterion. Reinforced concrete cores have been placed in the center sections of adobe walls (e.g., Petaluma Adobe, Sonoma State Historic Park). Cages of concrete beams, grade beams, and reinforced concrete columns have also been used (e.g., Plaza Hotel, San Juan Bautista, Calif.; and Cooper-Molera Adobe, Monterey, Calif.). These types of seismic retrofits are expensive and more intrusive than permitted by conservation standards.

In addition, the combination of concrete and adobe may result in problems of material compatibility that will be realized only years after the original retrofit.

Reinforced concrete bond beams placed at the tops of walls below the roof are often recommended for the upgrading of existing adobe buildings (State Historic Building Code 1990). The function of bond beams is to provide lateral support and continuity, though installation usually requires the removal of the roof system, which is an invasive and destructive procedure. Design is often based on elastic design criteria, resulting in a stiff bond beam. After cracks in adobe walls develop during an earthquake, the stiffness of the bond beam may exceed the stiffness of the walls by two or three orders of magnitude. Adobe walls have been observed to pull out from underneath such beams during an earthquake because of the difference in stiffness between the bond beam and the cracked wall sections, and the lack of a positive connection between the beam and the adobe walls.

Seismic upgrading of existing hazardous buildings has focused on the provision of maximum life safety to occupants, not on limitation of damage to the buildings. To date, the development of seismic upgrading practices has focused on multistory, unreinforced brick masonry buildings, a ubiquitous building type uniformly judged to pose the greatest life-safety hazard of all widely used building types in the United States. On first examination, unreinforced brick masonry structures might be considered to be similar to adobe: both are made of stacks of brick (masonry bricks are fired; adobe bricks are air-dried) formed into walls by joining the bricks with mortar. Adobe and adobe mortar are much weaker materials than brick and cement mortar; therefore, damage occurs at a much lower level of ground motion. More important, the walls of adobe buildings typically have a numerically smaller height-to-thickness ratio than the walls of brick buildings. This gives a different character to the stability problems for adobe than for brick buildings with their comparatively thin walls. Such differences should be accommodated in the seismic retrofit approaches used for the two materials.

Structural stability is fundamental for the adequate performance of adobe buildings during major earthquakes and for designing appropriate retrofit measures. The walls of adobe buildings will crack during moderate to large earthquakes because adobe walls are massive and both adobe brick and adobe mortar are low-strength materials. The walls have relatively little strength to resist the large inertial forces that are created within them during the ground accelerations of a seismic event. After cracks have developed, it is essential that the cracked elements of the structure remain stable, upright, and able to carry vertical loads.

A stability-based, retrofit design approach attempts to mobilize adobe's favorable postcracking, energy-dissipation characteristics while limiting relative displacements between adjacent cracked blocks. The GSAP investigations demonstrated that the stability-based approach can be the most effective method for providing life safety and for limiting the amount of damage during moderate to major earthquakes. The purpose of this approach is to prevent severe structural damage and collapse. Its proper application recognizes the limitations of adobe while taking advantage of the beneficial, inherent structural characteristics of historic adobe buildings. Thick adobe walls are inherently stable and have great potential for absorbing energy. These stability and energy-absorption characteristics can be enhanced by the application of a number of simple seismic improvement techniques, as described in the following sections.

Project Approach

GSAP adopted a phased project implementation: Phase 1 involved the evaluation of existing knowledge and practice and the development of interim technical guidelines for use in seismic strengthening of adobe structures. During phase 2, the necessary research was performed to develop an acceptable retrofit technology and to supplement what was currently known. Research included shaking-table tests as well as analytical modeling. In phase 3, a set of planning and engineering guidelines was drafted for the seismic stabilization of historic adobe structures, based on research results and professional judgment.

From the outset, a basic premise of the GSAP was that the guidelines produced as a culmination of the project would have the wide professional support of the technical community, not just the technical opinions of the few, and also be workable in application and responsive to real seismic retrofit problems. These practical principles governed the decision to approach GSAP as a cooperative endeavor of a wide group of individuals, including not only those who were expert in adobe seismic behavior but also others who were expert in all the issues concerned with the seismic improvement of historic buildings. Although it can be argued that it would have been less expensive and have taken less time to engage one person or firm to write a standard, it was felt that acceptance of such a standard would meet resistance throughout the professions and would probably not be widely used.

A GSAP Advisory Committee (see p. xviii) was also formed to review the project on a regular basis and to ensure that it was proceeding in a logical way to achieve its objectives. The GSAP Advisory Committee had two principal responsibilities: (1) to monitor project activities and advise the project manager on the management and direction of GSAP, and (2) to review the technical activities and accomplishments of GSAP and advise the project director and the project manager on its findings.

The advisory committee was appointed for the duration of the project. It met twice during the start-up phase and contributed regularly and substantially to the development of the work and research plans.

First-year activities

The first year of GSAP activities was initiated in 1991. During this period, the goal was to establish the groundwork on which research and guidelines development could be based. Detailed coverage of the first year's activities is contained in Thiel et al. (1991), which includes the following:

1. Preliminary assessment of engineering issues for development of the retrofit guidelines
2. Survey and assessment of selected California adobes to familiarize the GSAP team with the nature and problems of historic adobe buildings in context
3. Evaluation of the conservation principles involved in the seismic retrofit of culturally significant (historic) adobe structures
4. Review of the engineering and conservation characteristics of selected adobe seismic retrofits to determine the nature of the technical and conservation problems encountered in practice
5. Assessment of activities and approaches likely to encourage use of GSAP results
6. Glossary of terms used to describe historic adobes
7. Annotated bibliography of materials on adobe, adobe seismic performance, and adobe conservation
8. Inventory of historic California adobes

Second-year activities

The second year of GSAP was restricted to two principal activities:

1. Development of a preliminary draft of a planning guide for the seismic retrofit of historic adobe buildings
2. Design, performance, reporting, and assessment of shaking-table tests on models 1–3 to assess the effectiveness of stability-based retrofit measures

These activities constitute the major part of the second-year report (Tolles et al. 1993).

Third-year activities

The final year of GSAP activities consisted of the following:

1. Performance of shaking-table tests on small-scale models 4–9
2. Performance of shaking-table tests on large-scale models 10 and 11
3. Preparation of a survey on the damage to historic adobe buildings caused by the January 17, 1994, Northridge earthquake (Tolles et al. 1996)
4. Completion of the final project report, culminating in the present publication
5. Completion of the planning and engineering guidelines (Tolles, Kimbro, and Ginell n.d.)

Shaking-table tests on small-scale models
The purpose of the shaking-table tests was to study further the performance of adobe walls. Six small-scale (1:5) models were investigated during the final phase of the research effort. Three of the models had rectangular plans with no roof or floor system. The remaining three models were more complete. The design of the structures was based on the typical *tapanco*-style adobe buildings, which are characterized by gable-end walls, an attic floor, and roof framing. Buildings of this type are common throughout California.

Shaking-table tests on large-scale models
To investigate the possible effects of gravity loading on the effectiveness of retrofit measures and the patterns of damage, two large-scale (1:2) tapanco-style models were built and tested on a large shaking table. These models were instrumented to allow the measurement of building element displacements and stresses in the retrofit measures.

Survey of damage after the Northridge earthquake
The 1994 Northridge earthquake offered a rare opportunity to observe the tremendous loss that could occur to the limited numbers of historic adobe buildings remaining in California. It was a dramatic reminder of the need for the GSAP research and provided a useful educational tool for understanding the performance of historic adobe buildings after actual seismic events. The combination of field data collected following this earthquake (Tolles et al. 1996) and the results of laboratory research on small- and large-scale models, as described here, has led to a better general understanding of the seismic performance of adobe buildings.

Planning and engineering guidelines
A set of planning and engineering guidelines was drafted as a culminating activity of GSAP and will be published under the title *Planning and Engineering Guidelines for the Seismic Stabilization of Historic Adobe Structures* (Tolles, Kimbro, and Ginell n.d.). It contains two related parts: (1) a planning guide, which discusses the reasons for retrofitting historic adobe buildings, offers guidance on collecting important background information, reviews conservation principles as they apply to historic adobe buildings, and provides practical advice on planning seismic retrofits for historic adobe structures; and (2) an engineering guide, which provides information on designing seismic retrofits for historic adobe buildings, including general background on the seismic performance of adobes, information on both global and detail design, tools that may be used for retrofitting, as well as commonly observed types of damage and the recommended application of retrofit techniques for each type.

Chapter 2

Overview and Procedures for Models 1–9

The seismic testing research performed during the third year of the GSAP program was designed to further the understanding of the dynamic performance of adobe buildings and changes in that performance when various retrofit measures are applied. The experimental design was based on the present state of knowledge and was intended to increase the base of information on the seismic performance of adobe buildings.

The theoretical premise of this testing program was that the critical features of the seismic performance of historic adobe buildings will be evident after cracks in the building walls have fully developed. Because the walls are thick, cracking does not necessarily result in instability of the structure or its individual elements. Retrofit measures significantly improve seismic performance when they provide overall structural continuity, prevent instability, and provide restraint to reduce the relative displacements of cracked wall sections.

Six small-scale model buildings (models 4–9) were tested during the final phase of the GSAP research program. The first three models were similar to those tested during the second year of GSAP. The last three were more complete, tapanco-style models with gable-end walls, attic floor framing, and a roof system. Three model buildings (models 1–3) were tested during the second year of research (Tolles et al. 1993). A summary of the data on models 1–9 and their retrofits is listed in table 2.1.

Seismic Retrofit Techniques

The retrofit strategies selected for this testing program were chosen largely because of their potential for minimizing the post-elastic movements of cracked adobe blocks and their minimal impact on the historic fabric of the building. The selection criteria covered a much broader area than simply the effect on post-elastic performance, however. Criteria used for evaluating the strategies included the following:

1. Minimum effect on historic fabric of the structure and reversibility of the retrofit measures
2. Applicability of solutions appropriate to present building conditions

Table 2.1

Description of Getty Seismic Adobe Project (GSAP) small-scale model buildings

Model no.	S_L[a]	Date of tests	Walls	Type and location of retrofit
GSAP: Second-year research program				
Simple model: Four walls with no roof system[b]				
1	7.5	January 1993	NE	Upper horizontal strap
			SW	Upper and lower horizontal straps
2	7.5	January 1993	NE	Bond beam and center cores
			SW	Bond beam plus vertical and horizontal straps
3	7.5	January 1993	NE	Bond beam, center cores, and saw cuts; lower horizontal strap only in west pier of north wall
			SW	Bond beam, center cores, and internal lower horizontal straps
GSAP: Third-year research program				
Simple model: Four walls with no roof system				
4	5	January 1994	NE	Upper strap
			SW	Upper and lower straps
5	11	January 1994	NE	Control model, not retrofitted
			SW	Control model, not retrofitted
6	11	January 1994	NE	Bond beam, lower horizontal straps, and vertical straps
			SW	Bond beam, lower horizontal strap, and local ties
Tapanco-style model: Gable-end walls with attic floor and roof system				
7	5	September 1994	NW	Partial diaphragms applied on attic-floor and roof framing; upper and lower horizontal and vertical straps
			SE	Same as the NW walls, except vertical straps placed only on the piers between the door and window of the north wall
8	7.5	May 1995	NE	Partial diaphragms applied on attic-floor and roof framing; upper and lower horizontal and vertical straps
			SW	Partial diaphragms applied on attic-floor and roof framing; upper and lower horizontal straps and vertical straps; no lower horizontal strap on west wall; center-core rods
9	7.5	January 1994	NE	Control model, not retrofitted
			SW	Control model, not retrofitted

[a] S_L is the height-to-thickness (slenderness) ratio of the walls. For models that have more than one story (models 7–9), the slenderness ratio is the height of the wall from the foundation to the attic-floor framing.

[b] Model previously tested at Stanford University was used as the control for models 1–3 (Tolles and Krawinkler 1989).

3. Effectiveness in reduction of severe building damage and life-safety risks
4. Effectiveness in reduction of damage during moderate to severe events
5. Cost of retrofit and difficulty of installation
6. Use of retrofits for rapid installation in the stabilization of earthquake-damaged buildings

Any of the measures suggested must have a minimal impact on the building's historic fabric, and the solutions must be appropriate for the type of structural systems and for the conditions of these structures observed during the site survey. Clearly, any seismic retrofit measure should minimize risks to life and limb.

Certain solutions may satisfy life-safety requirements, but the structure may suffer severe damage during a major seismic event. Therefore, the fourth criterion relates to the reduction in the amount of damage suffered during moderate to major seismic events. In the evaluation of any retrofit, cost and difficulty of installation must be considered. Finally, it is important to develop retrofit techniques that can be safely used to stabilize earthquake-damaged structures in the time period immediately following damaging earthquakes.

From both a life-safety perspective and a conservation perspective, it is essential to prevent the collapse of buildings during major seismic events. Because adobe buildings typically have thick walls, the deflections they can undergo after cracks have fully developed are very great compared with the deflections that occur at the point when the building initially cracks. This is particularly true for the out-of-plane motions of these thick walls.

From a conservation perspective, it may also be important to evaluate ways of reducing intermediate levels of damage and to ensure that a building remains reparable. During a major seismic event, damage to a building may not threaten the life safety of its occupants, but the building may suffer substantial and irreversible damage. Additional retrofit measures may be added to improve the performance of a particular measure. For example, several measures used in model 8 were designed to minimize the extent of damage during very strong ground motions.

Overview of Tests

The objectives of the tests on models 1–3 were to demonstrate the effectiveness of certain stability-based retrofit measures. The objectives of the next round of tests (models 4–6) were to expand the knowledge of the seismic performance of adobe walls to include the effect of wall thickness. Tests on models 7–9 were designed to study more complete structures, such as the tapanco-style building. Models 1–6 were designed primarily to study the behavior of individual walls both in plane and out of plane, whereas models 7–9 were constructed to simulate the global behavior of a complete building system.

The results of the three 1:5 scale-model tests of adobe structures (models 1–3) demonstrated that the use of stability-based retrofit measures, which provide continuity and inhibit the relative displacements of cracked wall sections, can significantly improve the performance of an adobe building. The slenderness ratio (S_L)—the ratio of the height of a wall to its thickness—in these models was 7.5 (see Tolles et al. 1993). (The shorter the height, the lower the ratio. A thick-walled adobe building would typically have an S_L of about 5, whereas a thin-walled adobe might have an S_L of 10 or 11.)

The tests on models 4, 5, and 6 were performed to determine the effects of wall thickness on seismic behavior. Model 4 ($S_L = 5$) was designed with a retrofit similar to that of model 1 ($S_L = 7.5$). Model 5 ($S_L = 11$) was the unretrofitted control model for model 6 ($S_L = 11$), which was tested with retrofit measures similar to those used in model 2 (S_L = 7.5). Although there were some important differences in the behavior of models with different slenderness ratios, the specific retrofit system used was the dominant factor in the performance of these models. Wall thickness in the tested range turned out to be a secondary factor that influenced model behavior significantly only during very strong ground motions. In general, thicker walls improved the post-elastic performance because of the stability inherent in massive, low-to-the-ground walls.

Models 7–9 were more complete buildings. These models were constructed with the tapanco-style building elements. Model 7 (S_L = 5) was tested first and demonstrated the effectiveness of the retrofit system. Models 8 and 9 were constructed with thinner walls (S_L = 7.5) than those of model 7. The retrofit system used on model 8 was similar to that of model 7, except fiberglass center-core rods were installed in two adjacent walls. Model 9 was the unretrofitted control model used for comparison with the performance of model 8. In general, the retrofit systems used in models 7 and 8 were extremely effective and greatly improved overall seismic response.

Models 10 (unretrofitted) and 11 (retrofitted) were basically of the same design as tapanco models 9 and 8, respectively, except the model scale was increased from 1:5 to 1:2. The objective of these tests was to determine if gravity loads would affect the nature of the in-plane and out-of-plane wall motions and to assess the effectiveness of the retrofits in minimizing damage.

Description of Materials and Models

Adobe material

The adobe bricks and mortar for the models were made from a 1:5 mixture of clay and sand. These materials and the bricks' manufacture were chosen for their similarity to those used in previous Stanford testing (Tolles and Krawinkler 1989) so complementary results could be obtained. The clay was a pulverized, commercially available material that was mixed with well-graded silica and placed wet into molds. The molds were later removed and the bricks allowed to air-dry.

The brick-strength parameters were designed to be similar to those of an "average" adobe material found in Mexico, as in the earlier Stanford tests. During the previous test program, tests were performed to determine (1) flexure and compressive strength of the adobe material; (2) flexure of the individual bricks; and (3) flexure, compression, and diagonal tension of the brick assemblies. The compressive strength of the model assemblies was less than that of the prototype, but the more important flexural and diagonal tension properties were nearly the same.

Model design and construction

The walls of each model were one wythe (a single adobe brick) thick and constructed with a running bond. The length of the bricks was 7.9 cm (3.1 in.), the width dependent on the thickness of the walls, and the thickness of the bricks was 2.3 cm (0.9 in.). In the thin-walled models (models 5 and 6), the bricks were 5.3 cm (2.1 in.) wide, and the bricks in the walls of models 1–3 and models 8 and 9 were 7.9 cm (3.1 in.) wide. The bricks in the thick walls of models 4 and 7 were 11.7 cm (4.6 in.) wide. The mortar was made of the same clay:sand mixture as the adobe bricks.

The models were constructed on one of three concrete bases. After construction and drying for a minimum of 30 days, each model was transported to the shaking table. Steel dowels, cast into the concrete base, were projected into the first two courses of adobe. These dowels were used to limit slipping of the model along the base; slippage at the base would be expected in these models because the vertical loading was not fully simulated in the "gravity forces neglected" models used in this research project. Wood lintels, the thickness of one course of adobe, were used over all door and window openings.

Retrofit measures

The retrofit measures on the model buildings included horizontal elements and, usually, vertical elements. The horizontal elements were either nylon straps, a bond beam, or a partial wood diaphragm. The vertical elements were either nylon straps, center-core rods, or local ties.

The wood bond beams used on models 3 and 6 were made of Douglas fir and were 3.8 cm (1.5 in.) wide and 1.0 cm (0.375 in.) thick. The bond beams were anchored to the walls with 0.3 cm (0.125 in.) diameter by 8.9 cm (3.5 in.) long coarse-threaded screws. The holes for the screws were predrilled before placement.

The vertical and horizontal straps were made of a 0.3 cm (0.125 in.) wide, flexible, woven nylon strap typically used for a bootlace. The straps always formed a loop either around the entire building or around an individual wall. The straps were passed through small holes in the wall and the two ends were knotted together. Exterior straps were found to have been useful for stabilizing some Guatemalan adobes following the major earthquake in that region in 1976 (Molino de Garcia 1990).

Crossties were made of 0.16 cm (0.062 in.) diameter nylon cord and were installed to reduce the differential displacement across cracks. Also, when vertical and horizontal straps were installed on both sides of a wall, crossties were added to provide a through-wall connection. The crossties were inserted through small holes in the wall to greatly reduce the displacement that could occur perpendicular to the plane of the wall. Flat nylon straps that are commonly used in electrical work and referred to as cable ties were also used as crossties.

Although stresses in the crossties and vertical or horizontal straps were not measured on the 1:5 scale models, none of the straps or crossties failed during any of the tests. The static breaking load of the nylon straps was 102 kg (225 lb.) and the breaking load of the cable ties was 27 kg (60 lb.).

The center-core elements used in models 2 and 3 were 0.3 cm (0.125 in.) diameter steel drill rods. The rods were drilled directly into the adobe after flattening each end into a V-shaped form. The rods, which were left in place after drilling, functioned well throughout the testing sequence and were adequate for providing shear dowels between the cracked sections of the walls.

The center-core elements in model 8 were 0.48 cm (0.188 in.) diameter steel rods anchored with an epoxy grout. The holes were drilled with a 0.6 cm (0.25 in.) diameter drill bit, but, given the coarseness of the sand in the adobe mixture, the actual diameter of the holes was approximately 1.0 cm (0.375 in.). After the testing, one rod was removed from the wall, and the epoxy core was found to be nearly 1.3 cm (0.5 in.) in diameter. Essentially, then, this was the effective center-core diameter.

All center-core rods were located entirely within the adobe wall and were not connected to the concrete base. When used in conjunction with wood bond beams, the rods were anchored to the bond beam with an epoxy resin.

Model similitude

Modeling theory establishes the rules by which the geometry, material properties, initial conditions, and boundary conditions of the model and the prototype can be related. The laws of similitude for linear elastic behavior are based on well-established principles of dimensional analysis and lead to the development of a complete set of correlation functions (scaling laws) that define the model-prototype relationship. A listing of similitude requirements is given in table 2.2.

Table 2.2
Similitude requirements

Model scaling parameters[a]		**Model type**		
		True replica	Artificial mass simulation	Gravity forces neglected
Length	l_r	l_r	l_r	l_r
Time	t_r	$l_r^{1/2}$	$l_r^{1/2}$	$l_r = \left(\frac{E}{\rho}\right)_r^{-1/2}$
Frequency	ω_r	$l_r^{-1/2}$	$l_r^{-1/2}$	$l_r^{-1}\left(\frac{E}{\rho}\right)_r^{1/2}$
Velocity	v_r	$l_r^{1/2}$	$l_r^{1/2}$	$\left(\frac{E}{\rho}\right)_r^{1/2}$
Gravitational acceleration	g_r	1	1	neglected[b]
Acceleration	a_r	1	1	$l_r^{-1}\left(\frac{E}{\rho}\right)_r$
Mass density	ρ_r	E_r/l_r	augmented[c]	ρ_r
Strain	ϵ_r	1	1	1
Stress	σ_r	E_r	E_r	E_r
Modulus of elasticity	E_r	E_r	E_r	E_r
Displacement	δ_r	l_r	l_r	l_r
Specific stiffness	$\left(\frac{E}{\rho}\right)_r$	l_r	augmented	$\left(\frac{E}{\rho}\right)_r$
Force	F_r	$E_r l_r^2$	$E_r l_r^2$	$E_r l_r^2$

From Moncarz and Krawinkler 1981.
[a]Subscript notation: m = model; p = prototype; r = ratio between model and prototype (e.g., $l_r = \frac{l_m}{l_p}$).
[b]Effects of gravity are neglected; this modeling theory assumes that the effects of gravity forces are minor and are negligible.
[c]Mass of the building augmented by adding additional, structurally ineffective mass to the building.

The models used in this study are referred to as "models without the simulation of gravity loads." The same type of model was used in previous Stanford research. The model-prototype relationship is clearly defined during the time when the adobe material is still in the elastic range. This type of model is accurate up to the point at which cracks develop because the vertical loads are small and have minor effects on the elastic stresses in walls of single-story structures.

When a building is damaged and becomes inelastic, the accuracy of the model is more difficult to assess. Overturning of individual walls is not properly modeled because gravity forces resist overturning. Also, sliding along cracks is not accurately simulated in the "gravity forces neglected" model, shown in table 2.2. Resistance to sliding is directly proportional to the vertical stresses, which are smaller because gravity loads are not fully simulated. Such resistance is also affected by the increased frequency characteristics of the model and by input motions (Tolles and Krawinkler 1989).

The problems discussed here were not considered to be of primary importance since the purpose of research on the small-scale models was to study the global response characteristics of the models and to evaluate the relative merit of the different retrofit measures. Because the linear elastic modeling of a single-story adobe building is nearly exact, the crack patterns should be nearly the same as those found in the prototype. The response of the cracked models will contain the global characteristics of the prototype, even though overturning will occur at lower levels and frictional resistance along cracks may be lower.

A roof system was not included on models 1–6 because the primary purpose of these tests was to study the in-plane and out-of-plane performance of individual walls. Performance could be studied more independently by not including a roof system that would directly couple the load-bearing walls. If a roof system had been included, it might have affected the behavior of the models.

A roof system was included in models 7–9, however, to determine the extent of any behavioral differences. The 1:5 scale and the slenderness ratio were chosen because they were identical to six models tested at Stanford University in the mid-1980s (Tolles and Krawinkler 1989).

Description of Test Procedure

Test setup

The small-scale models were tested on the shaking table at the John A. Blume Earthquake Engineering Center at Stanford University. The shaking table is 1.52 m (5.0 ft.) square and has a uniaxial motion with maximum displacements of 7.6 cm (3 in.). Because the scale of the models was 1:5, the maximum displacements in the prototype domain were 38 cm (15 in.). Displacement capacity is a critical feature in determining a shaking table's ability to cause out-of-plane collapse of thick-walled buildings.

Each model was built on a 1.67 m (5.5 ft.) square concrete slab and allowed to dry for a minimum of 30 days. The models were transported and secured to the shaking table the day before testing began. Four 30 cm (12 in.) square plywood blocks were placed on the shaking table. Cement mortar was placed between each block and the concrete slab to allow for leveling and equal contact at each block. After the mortar was allowed to cure overnight, the slab was anchored to the table using four 45 cm (18 in.) long by 2.2 cm (0.875 in.) diameter bolts attached to the top of the slab and anchored to two 10 cm (4 in.) square timbers placed on the underside of the shaking table.

Instrumentation

Instrumentation for these tests was kept to a minimum. The goal of the testing program was to maximize the number of tests that could be performed and to make a qualitative evaluation of the performance of the models based on visual observations. The displacements and accelerations of the shaking table were measured to verify the proper performance of the table and for comparison with previous testing. The displacements were measured using the linearly variable differential transformer (LVDT) that is part of the shaking table. Accelerations were measured at the top of the shaking table and at the top of the concrete slab to check for slippage between the table and slab. No other quantitative measurements were collected.

Simulated earthquake motions

The earthquake motion used in these tests was based on the N21E component of the 1952 Taft earthquake in Kern County, California. Each model was subjected to a series of ten simulated earthquake displacement motions. Each subsequent displacement motion was 20–30% larger than the previous one. The displacement, velocity, and acceleration records for the N21E component of the Taft earthquake are presented in figure 2.1. The maximum simulated earthquake motion was 6–7 times larger than the original earthquake, based on comparison of the displacement records.

A listing of the ten simulated earthquake motions is presented in table 2.3. The estimated peak ground acceleration (EPGA) is similar to the actual peak ground acceleration (PGA) except that the EPGA is determined from the response spectrum between 2 and 8 hertz (Tolles and Krawinkler 1989).

Test level is used to describe the intensity of a test in terms of peak displacement and acceleration. In most model tests, the sequence was linear from test level I to test level X. This was not true for all buildings. The test sequence for model 5 skipped from test level V to test level VIII in an attempt to accelerate the collapse. In many buildings, test level X was repeated more than once. In these cases, the first repetition is noted as test level X(1) and the second as test level X(2). Test level IX was repeated twice in the test sequence for model 8.

Test levels VIII, IX, and X represented very large ground motions. In the prototype domain, the peak horizontal displacement for

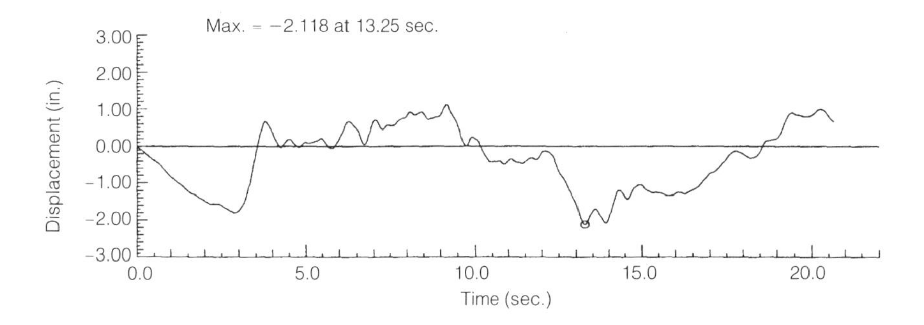

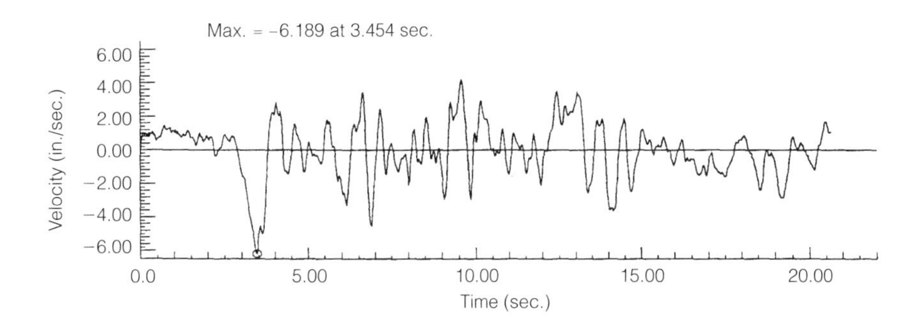

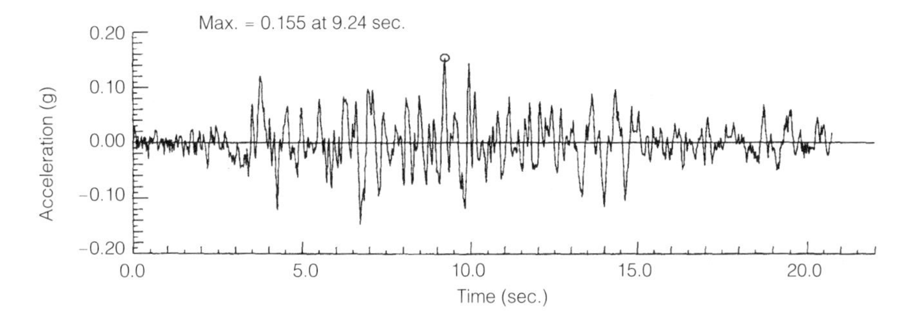

Figure 2.1
Graph of N21E component of 1952 Taft earthquake (displacement, velocity, and acceleration).

test level X was 38.1 cm (15 in.), and the peak acceleration was approximately 0.58 g. The peak accelerations for the higher test levels were not fully replicated by the table motions because of the shaking table's inability to represent stronger ground motions. Nevertheless, the higher frequency component of the table motion was not important at the higher test levels because of the damage that had already been sustained by the model buildings. Because peak accelerations are often used as a means of comparison, if the table had been capable of fully duplicating the earthquake, the peak acceleration for test level X would have been between 0.8 g and 0.9 g. Nevertheless, the Taft earthquake record has a substantial high-frequency component, and complete simulation of the high-frequency component during the larger table motions would probably not be accurate.

The higher level tests were at the upper end of what might be expected during the most severe earthquake in California. A dynamic displacement of 38.1 cm (15 in.) is very large but definitely possible. Even if the tests were carried out at a slightly greater acceleration than that experienced in the largest possible earthquake, they represent an effective means of testing the resilience of the proposed retrofit strategies.

Table 2.3
Simulated earthquake motions for testing (dimensions in the prototype domain)

Test level	Maximum EPGA[a] (g)	Maximum displacement cm	in.
I	0.12	2.54	1.00
II	0.18	5.08	2.00
III	0.23	7.62	3.00
IV	0.28	10.16	4.00
V	0.32	12.70	5.00
VI	0.40	15.88	6.25
VII	0.44	19.05	7.50
VIII	0.48	25.40	10.00
IX	0.54	31.75	12.50
X	0.58	38.10	15.00

[a]Estimated peak ground acceleration

In addition, the quality of laboratory-built models generally is higher than that found in the field, whether due to the original construction or to subsequent damage. Therefore, it was important to test these "high-quality" models with powerful earthquake forces.

The duration of the ground motion was 20 seconds in the prototype domain. This is a long duration for the lower test levels but a short duration for the higher levels. An earthquake of magnitude 8 on the Richter scale might last longer than a minute. Even though these model buildings were subjected to repeated ground motions, which caused cumulative damage to the models, the combined effect of tests VIII, IX, and X may be a reasonable representation of the largest expected ground motions in the California area.

The model buildings were subjected only to uniaxial motion. This shaking mode is preferred during tests that are used to isolate the dynamic characteristics of the models. All models were shaken in an east-west direction.

Documentation

The behavior of the models was documented using drawings, photographs, and videotape. The acceleration and displacement of each test were recorded digitally. Still photographs were taken of each wall of the model to record the damage after every test.

The dynamic motion of all model walls was recorded on videotape. Four video cameras aimed at each of the four wall intersections were used to record the motion of adjacent walls. The dynamic motion of these tests lasted a little longer than 4 seconds, which was equivalent to 20 seconds in the prototype domain. It was important to record the dynamic motions of the models from all angles to be able to study the wall motions after the testing was completed.

During a test sequence, the cracks that developed were numbered as they appeared. The number of each crack coincided with the number of the test in which the crack occurred.

Chapter 3
Description of Tests for Models 4–6

The purpose of the tests on models 4–6 was to investigate the effect of the slenderness ratio of the walls on the performance of the model buildings. Model 5 was unretrofitted, and models 4 and 6 were retrofitted. Because only one model was not retrofitted, the evaluation of the effect of wall thickness was based primarily on the performance of the retrofitted models.

Description of Models

Retrofit strategies used on models 1 and 4 were similar. The measures used on models 2 and 6 were also similar. Details of the tests and results on models 1–3 are given in the GSAP second-year report (Tolles et al. 1993:3.39–75).

The only difference between the retrofits of models 1 (S_L = 7.5) and 4 (S_L = 5.0) was the spacing of the crossties. A few additional ties were added to model 4 in areas where crack displacements had been large during the higher-level tests performed on model 1.

The differences between models 2 and 6 were more significant. Each model had a wood bond beam, and two walls of each model were retrofitted with vertical straps. In model 2, the east and north walls had steel center-core rods. In model 6, the south and west walls were fitted with only local crack ties, which were installed to evaluate the possibility of the use of a less invasive approach. For the most part, the use of local crack ties was not successful. In model 3, all walls contained center cores.

The general layout of the models is shown in figure 3.1. Each wall had a door and a window. The door was located near the center of the wall, and the windows were located close to the southwest and northeast corners. The same retrofit was used on two adjacent walls, allowing simultaneous evaluation of a specific retrofit on both in-plane and out-of-plane walls. Table motion was uniaxial in an east-west direction.

Model 4 (S_L = 5)

The retrofit on model 4 consisted of upper horizontal straps applied to both sides of all four walls and an additional lower horizontal strap on

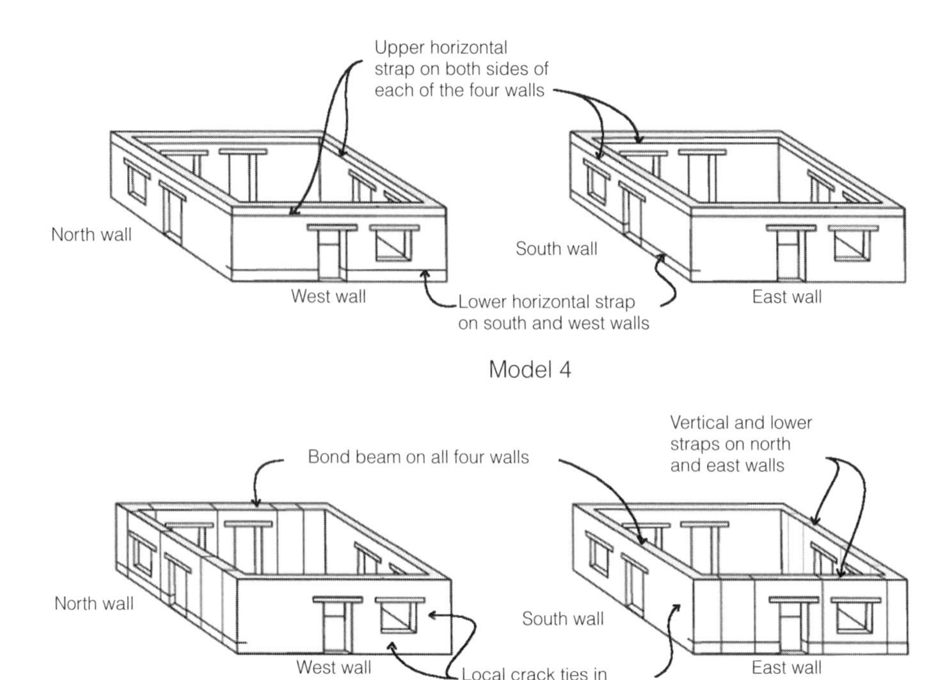

Figure 3.1
General layout of models 4 and 6, showing inner and outer wall straps connected through the wall by thin nylon cords at 15.2–20.3 cm (6–8 in.) intervals.

the south and west walls. The upper horizontal strap was located at approximately the midpoint between the top of the wall and the top of the door and window openings. The lower horizontal strap on the south and west walls was located at approximately two-thirds the distance from the foundation to the bottom of the windows.

Horizontal straps were located on both sides of each wall. Crossties were placed at points approximately 10 cm (4 in.) on center, and connected straps on opposite sides of the walls.

The retrofits installed on model 4 were the same as those on model 1. Model 1 performed through test level IX and collapsed during test level X. Model 4 was designed to determine whether the effectiveness of this retrofit strategy could be improved when used on a model with thicker walls.

Model 5 (S_L = 11)

Model 5 was used as an unretrofitted, thin-wall control model and had the same layout as models 1–4 and 6.

Model 6 (S_L = 11)

The retrofits used on model 6 were similar to those used on model 2; the north and east walls (see fig. 3.1) were retrofitted in an identical manner. A bond beam was applied at the top of all four walls in addition to the vertical straps and lower horizontal straps that were applied to two of the walls (the north and east walls in model 6 and the south and west walls in model 2). Small-diameter center-core rods were used in the north and east walls of model 2, but rods were not installed in model 6 because the walls were too thin to allow accurate placement. Instead, the south and west walls of model 6, which were more lightly retrofitted than the north and east walls, were retrofitted with local crack ties to connect the predicted major wall segments. Lower horizontal straps were used on this model. Local crack ties were used for connections across locations where cracks were anticipated.

Test Results for Model 4

Model 4 (S_L = 5) performed well throughout the testing sequence, although it nearly collapsed during test level X (table 3.1). The condition of the model following selected tests is shown in figures 3.2–3.7.

Damage began during test level IV when cracks developed primarily in the in-plane walls (fig. 3.2). The thick walls (S_L = 5) resisted the out-of-plane motions, but the in-plane forces exceeded the strength of the adobe.

The development of the primary system of cracks was completed during test levels V and VI (figs. 3.3 and 3.4). At this point in the test sequence, the retrofit system became active.

Although no thick-walled (S_L = 5) control model was tested, it would appear that an unretrofitted model (S_L = 5) would have collapsed during test level VII (fig. 3.5) or VIII. During these tests, the retrofit system was very active, and the cracked wall sections began to undergo permanent displacements.

Prior to test level IX, local crack ties were added to each of the piers between the doors and windows because of the permanent

Table 3.1
Model 4: Test sequence and commentary (prototype dimensions)

Test level	EPGA[a] (g)	Peak displacement cm	Peak displacement in.	Comments
III	0.23	7.62	3.00	No damage
IV	0.28	10.16	4.00	Crack initiation primarily in the in-plane walls
V	0.32	12.70	5.00	Further crack development
VI	0.40	15.88	6.25	Complete crack system development
VII	0.44	19.05	7.50	Retrofit system fully active
VIII	0.48	25.40	10.00	Permanent displacement of window-door pier blocks
IX	0.54	31.75	12.50	Local crack ties added to pier blocks; building system performs well
X	0.58	38.10	15.00	Near-collapse of east wall; substantial permanent displacements throughout the model

[a]Estimated peak ground acceleration

(a)

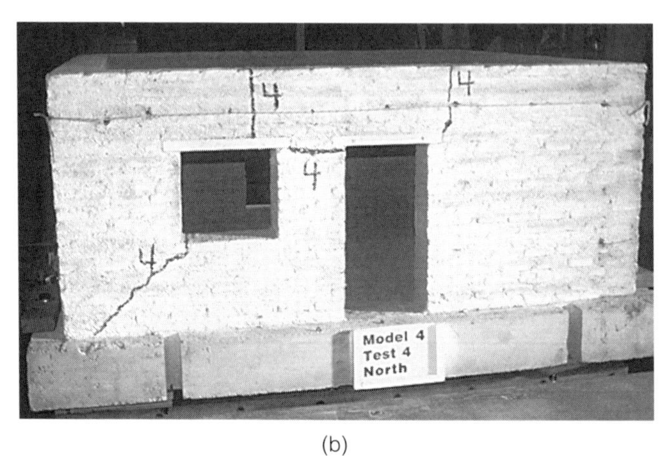

(b)

(c)

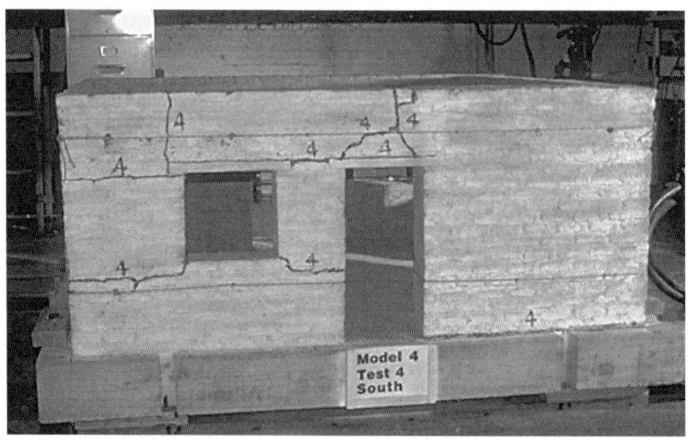

(d)

Figure 3.2
Model 4, test level IV, showing (a) east wall (out of plane); (b) north wall (in plane); (c) west wall (out of plane); and (d) south wall (in plane). Crack initiation occurred, with crack pattern in walls showing little differentiation between in-plane and out-of-plane directions. Cracks in the in-plane walls (north and south) started at the corners of openings and at the pier between door and window. In the out-of-plane direction, the west wall sustained crack damage in several locations, while the east wall had only one crack, extending diagonally from the bottom of the window toward the lower north corner. Note that in this thick-walled model, cracks began in the in-plane walls.

(a)

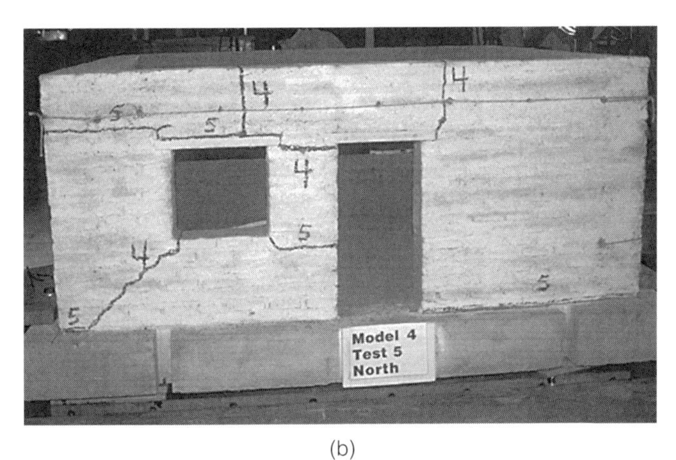

(b)

(c)

(d)

Figure 3.3
Model 4, test level V, showing (a) east wall (out of plane); (b) north wall (in plane); (c) west wall (out of plane); and (d) south wall (in plane). Some additional cracks were initiated in the in-plane walls and the east, out-of-plane, wall. No additional crack damage was observed in the west wall. At this point in the testing, dynamic displacements remained small and the retrofit system (horizontal straps) had little or no effect on the model's performance.

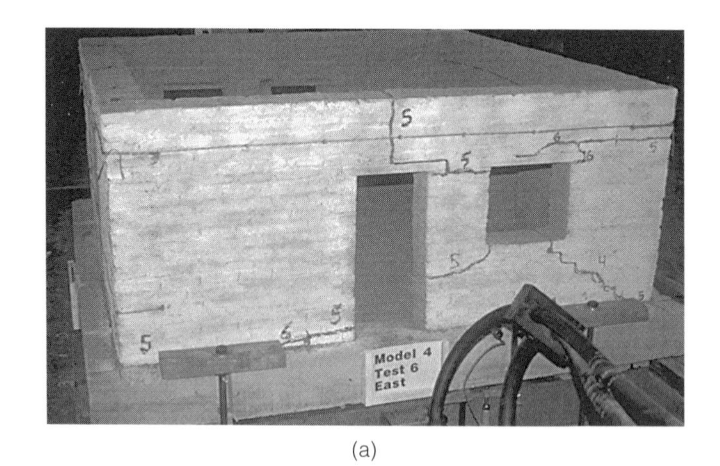

(a)

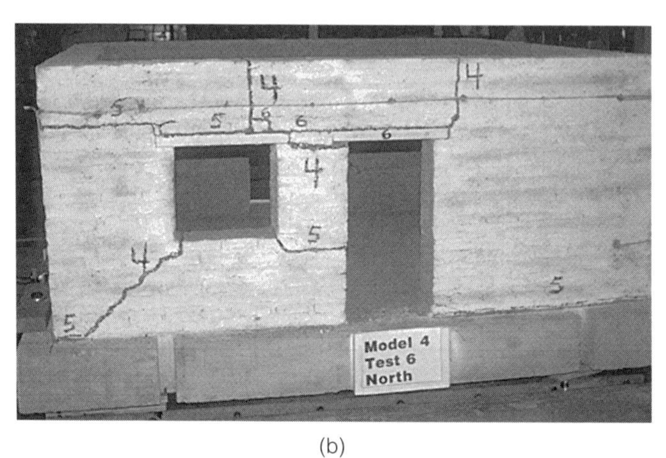

(b)

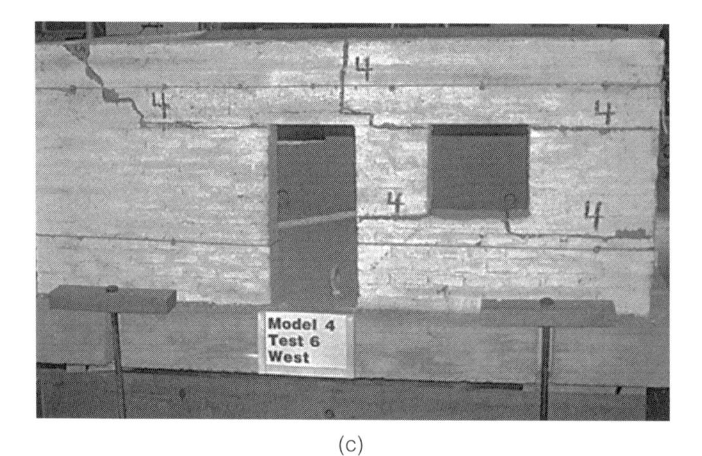

(c)

(d)

Figure 3.4
Model 4, test level VI, showing crack damage in (a) east wall (out of plane); (b) north wall (in plane); (c) west wall (out of plane); and (d) south wall (in plane). Although cracking remained nearly constant from that of test level V, the motion of the cracked wall sections became substantially larger during this test level as the cracks began to open up and the wall sections started to move relative to one another. There was little rocking of the cracked wall sections in the out-of-plane walls. The retrofit straps were engaged during this test sequence.

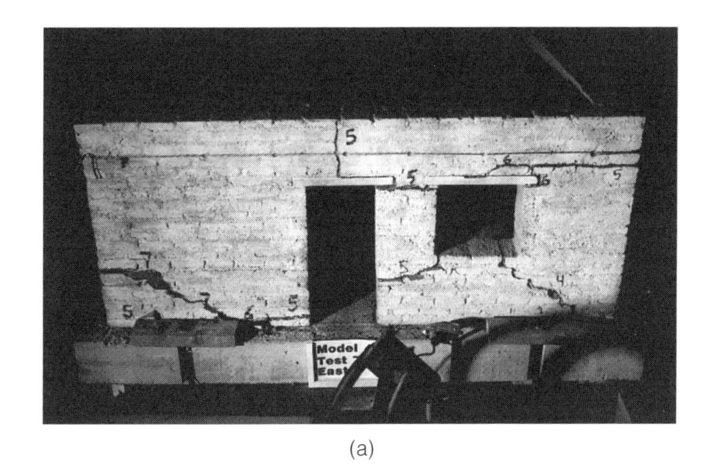

(a)

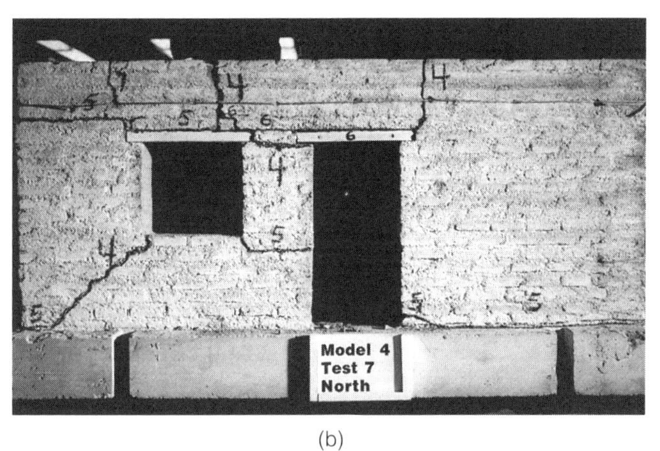

(b)

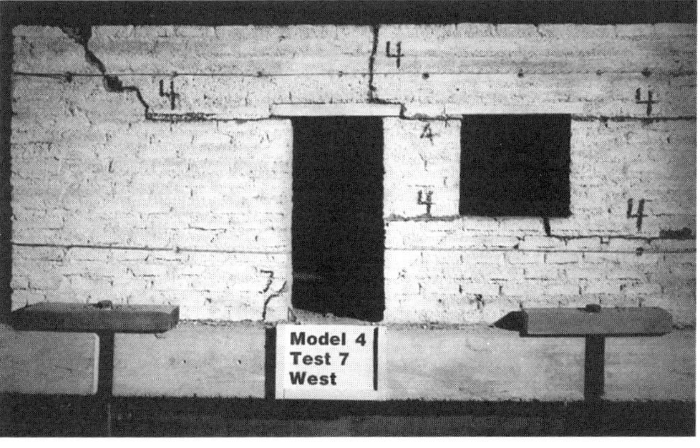

(c)

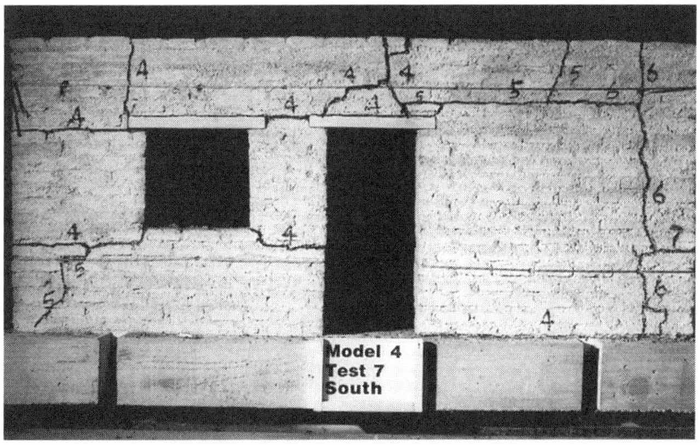

(d)

Figure 3.5
Model 4, test level VII, showing crack pattern in (a) east wall (out of plane); (b) north wall (in plane); (c) west wall (out of plane); and (d) south wall (in plane). Crack pattern was nearly complete. In the east wall, cracks developed along the base of the main panel as the wall rocked back and forth at the base. A similar crack extended completely across the base of the west wall. The retrofit straps were definitely effective during this test. They acted to restrain the out-of-plane motions of the walls and the relative in-plane displacements of the cracked wall sections. (Note: The photographs for test level VIII, which show only a few additional cracks and crack extensions, are not included here.)

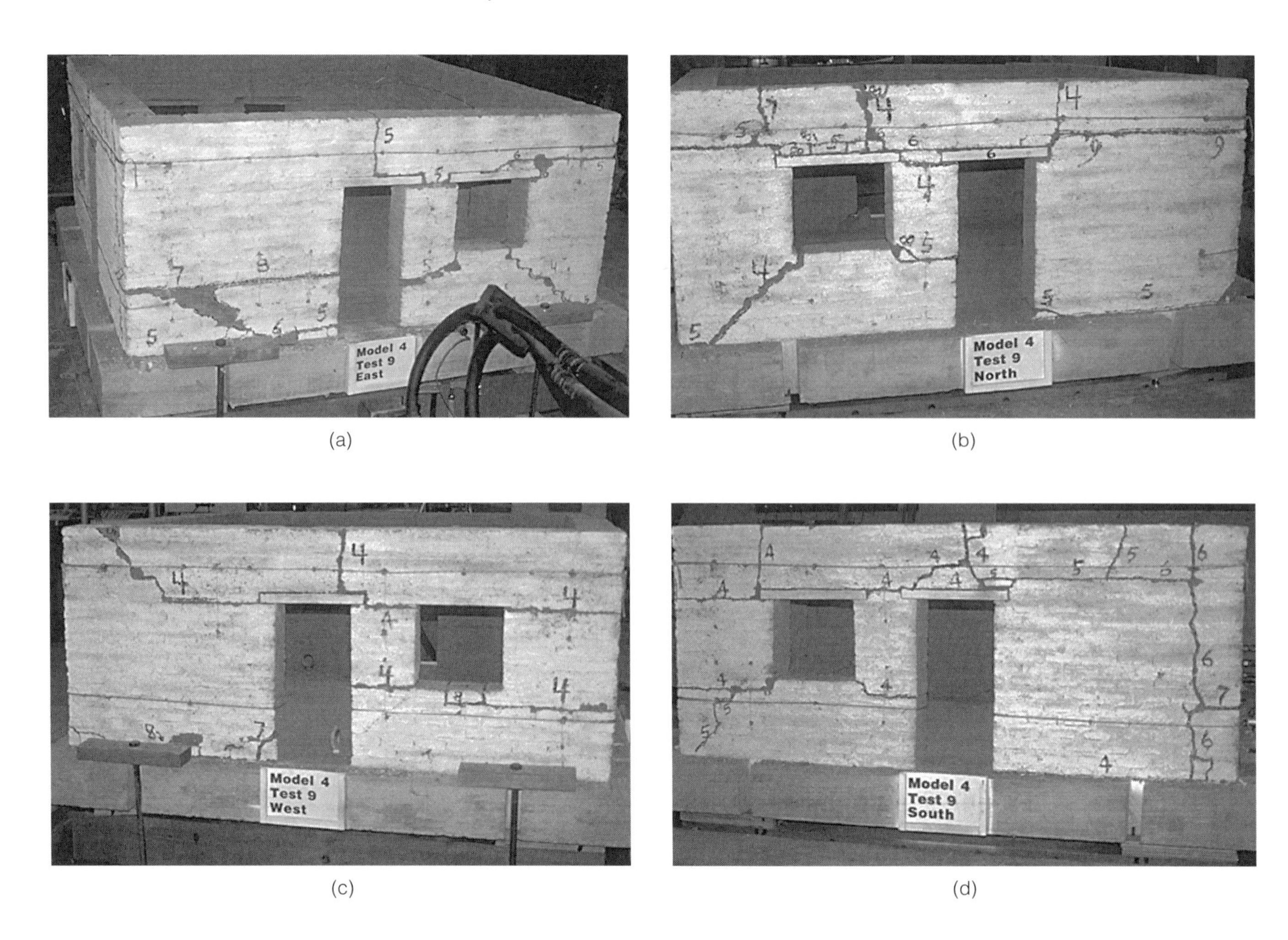

Figure 3.6
Model 4, test level IX, showing crack damage to (a) east wall (out of plane); (b) north wall (in plane); (c) west wall (out of plane); and (d) south wall (in plane). (Test level cracks are not marked on east and west walls; see Fig. 3.7a and c.) Before this test level, local crack ties were added to the pier blocks between the doors and windows because of the large displacements sustained during test level VIII. During test levels VIII and IX, rocking of the wall sections was greatly restrained by the upper horizontal strap. This restraint generated additional horizontal cracks in the large panel to the left of the door in the east wall at about one-third height.

offsets sustained during test level VIII. These blocks were pushed back into position before the ties were applied. The model behaved well during test level IX (fig. 3.6), although there were permanent offsets between many cracked wall sections.

The model nearly collapsed during test level X (fig. 3.7). The east wall was leaning outward at the end of the test. The wood lintel over the door of the east wall was dislodged, and the adobe block above it was nearly dislodged.

Test Results for Model 5

Model 5 was an unretrofitted control model, tested to demonstrate the vulnerability of unretrofitted adobe walls with $S_L = 11$. Cracks began to form during test level III (fig. 3.8), and the crack pattern fully developed during test level V (fig. 3.9) with the formation of flexural cracks in the out-of-plane walls and shear cracks in the in-plane walls (test level IV was omitted; table 3.2). Considerable rocking was observed during this test. The sequence was advanced to test level VIII, the fifth test in this series, which resulted in the complete collapse of the east wall and most of the west and south walls (fig. 3.10). The north wall was severely damaged, and the eastern corner was completely destroyed.

(a)

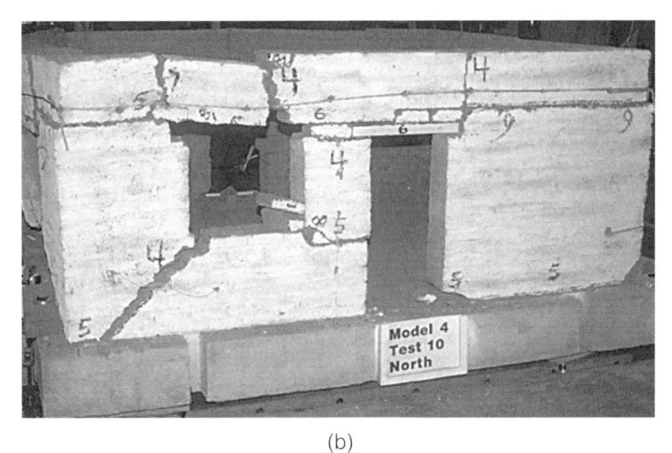

(b)

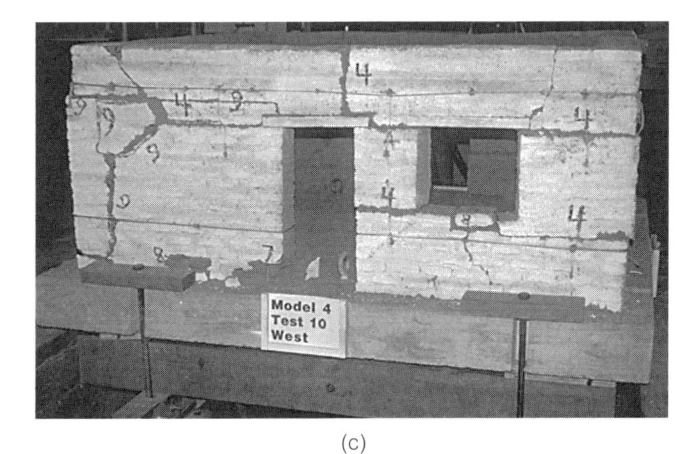

(c)

(d)

Figure 3.7
Model 4, test level X, showing damage to (a) east wall (out of plane); (b) north wall (in plane); (c) west wall (out of plane); and (d) south wall (in plane). The model nearly collapsed at this test level. The east wall was leaning outward, and the west wall would have fallen if not for the upper horizontal strap. The wood lintel over the window in the north wall was dislodged. A horizontal crack in the upper portion of the walls extended around the entire model. The piers between the door and window on both east and west walls were held in place by local ties that were added after test level IX.

(a)

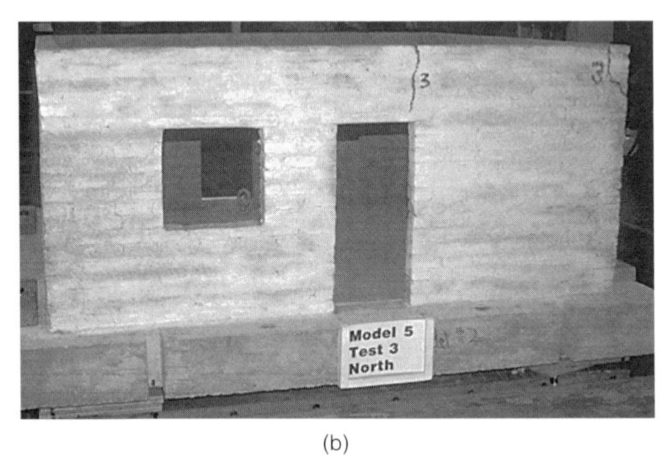

(b)

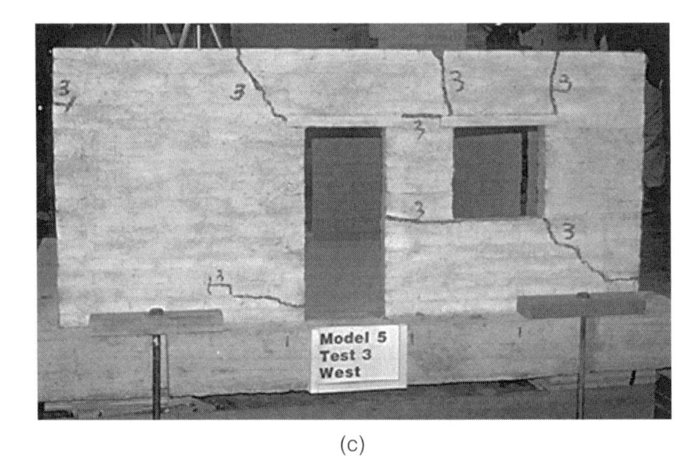

(c)

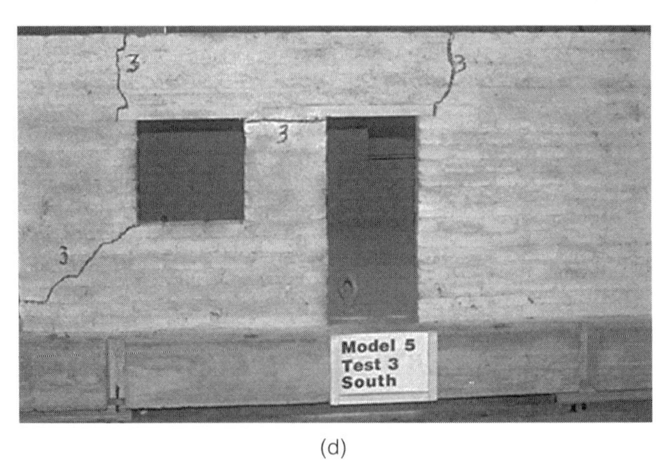

(d)

Figure 3.8
Model 5, test level III, showing (a) east wall (out of plane); (b) north wall (in plane); (c) west wall (out of plane); and (d) south wall (in plane). Minor cracking is seen with little visible out-of-plane wall displacement. Similar types and numbers of cracks developed in both in-plane and out-of-plane walls.

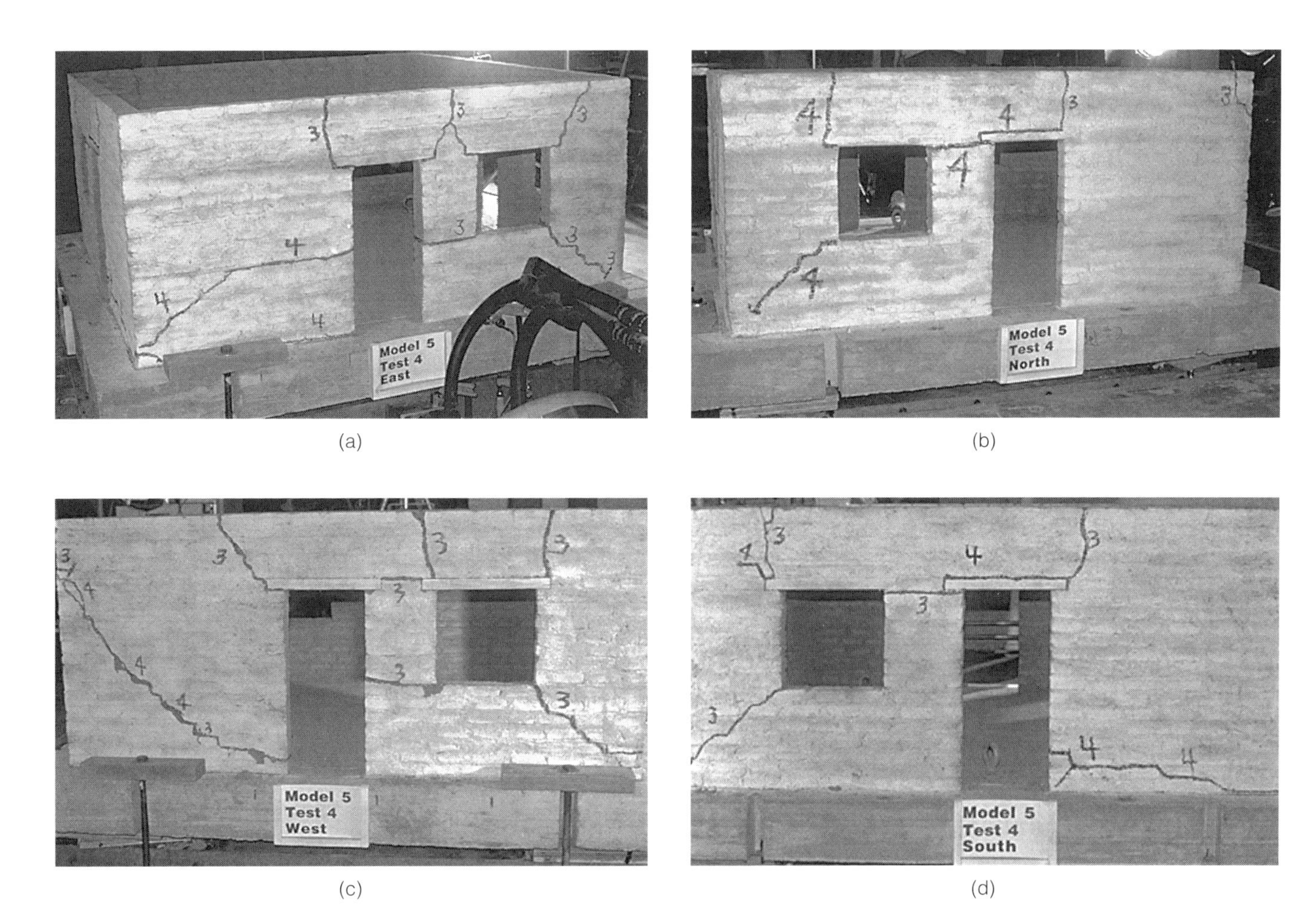

Figure 3.9
Model 5, test level V (fourth in the test sequence), showing cracking in (a) east wall (out of plane); (b) north wall (in plane); (c) west wall (out of plane); and (d) south wall (in plane). Fully developed crack pattern is visible with flexural cracks in the out-of-plane walls and shear cracks in the in-plane walls. The walls rocked noticeably during the test but did not collapse.

Test Results for Model 6

As expected, the performance of the retrofitted model 6 was much better than that of model 5. During test levels III–V, the crack patterns were similar to those of the unretrofitted model (table 3.3; figs. 3.11–3.13). Large flexural, yield-line cracks developed in the east and west walls, which demonstrated that the bond beam had little effect on the elastic behavior of the model. Few additional cracks developed between test levels IV and VI.

Table 3.2
Model 5 (control model): Test sequence and performance summary (prototype dimensions)

Test level	EPGA[a] (g)	Peak displacement cm	Peak displacement in.	Performance summary
I	0.12	2.54	1.00	No damage
II	0.18	5.08	2.00	No damage
III	0.23	7.62	3.00	Crack initiation
V	0.32	12.70	5.00	Full crack pattern development; near collapse
VIII	0.48	25.40	10.00	Collapse of 75% of the walls; complete collapse of out-of-plane walls

[a]Estimated peak ground acceleration

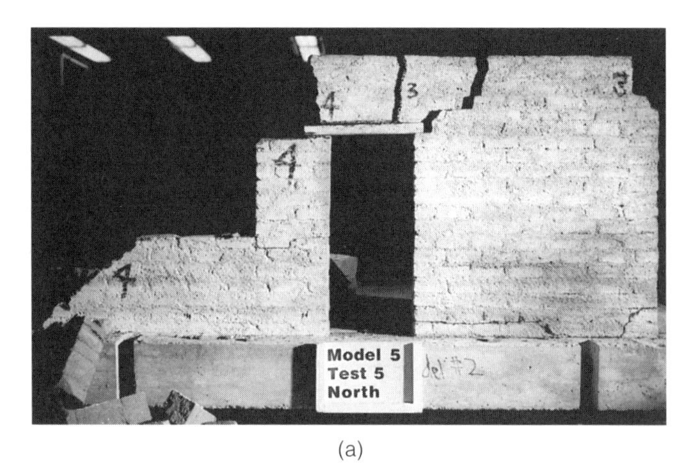

(a)

(b)

(c)

Figure 3.10
Model 5, test level VIII (fifth in the test sequence), showing results on (a) north wall (in plane); (b) south wall (out of plane); and (c) west wall (in plane). The model was almost completely destroyed, with only part of the north wall remaining erect.

During test levels VI and VII, the retrofit system was very active (figs. 3.13 and 3.14). The out-of-plane walls rotated about their bases, allowing substantial dynamic displacement of the east and west walls. The magnitude of these displacements was governed by the mass of the wall, coupled with the stiffness of the bond beam.

Table 3.3
Model 6: Test sequence and performance summary (prototype dimensions)

Test level	EPGA[a] (g)	Peak displacement cm	in.	Performance summary
III	0.23	7.62	3.00	Crack initiation
IV	0.28	10.16	4.00	Further crack development
V	0.32	12.70	5.00	Complete crack system; bond beam fully active with large out-of-plane displacements
VI	0.40	15.88	6.25	Bond beam fully active, allowing substantial out-of-plane displacements of east and west walls
VII	0.44	19.05	7.50	
VIII	0.48	25.40	10.00	Collapse of lightly retrofitted west wall
IX	0.54	31.75	12.50	Additional partial collapse of south wall
X	0.58	38.10	15.00	Partial collapse of east wall

[a]Estimated peak ground acceleration

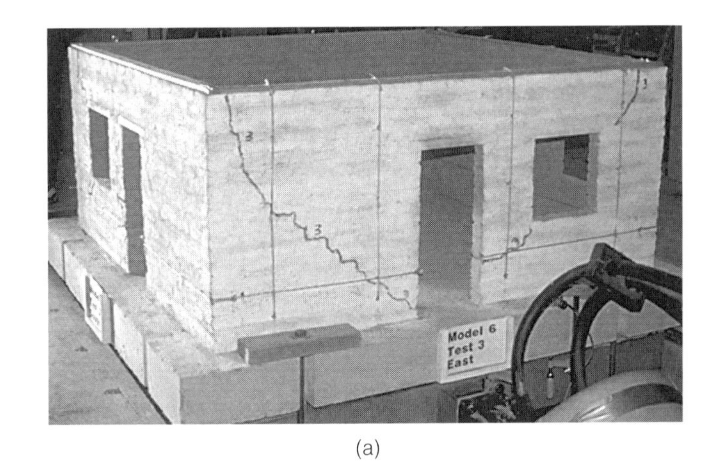

(a)

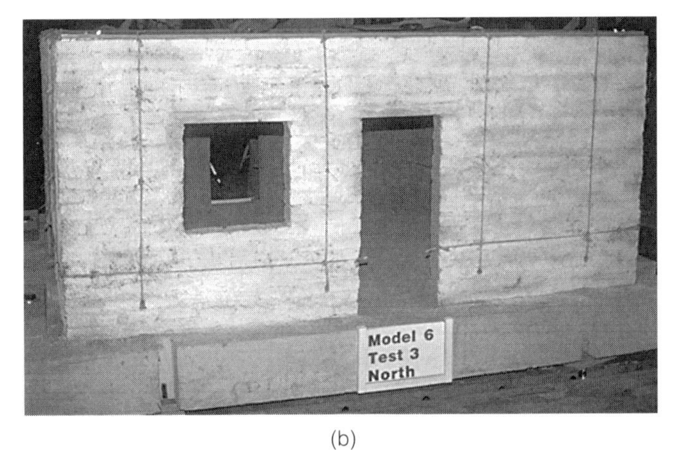

(b)

(c)

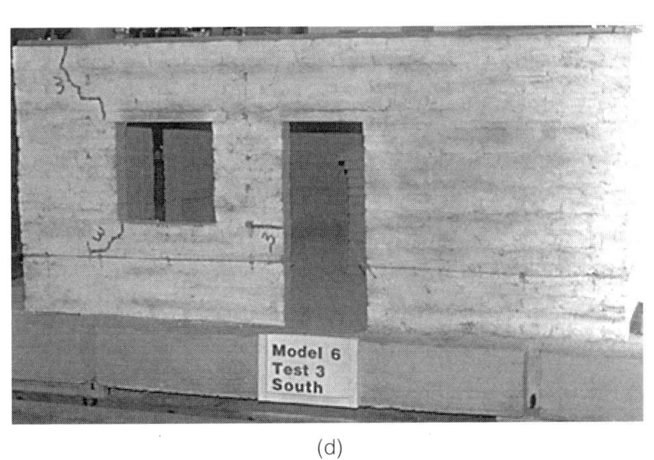

(d)

Figure 3.11
Model 6, test level III, showing (a) east wall (out of plane); (b) north wall (in plane); (c) west wall (out of plane); and (d) south wall (in plane). Initial cracking is visible, and a classic flexural pattern was observed in the east wall, though only small cracks were observed in the west wall. No cracks developed in the north wall, and only two or three slight cracks developed in the south wall.

(a)

(b)

(c)

(d)

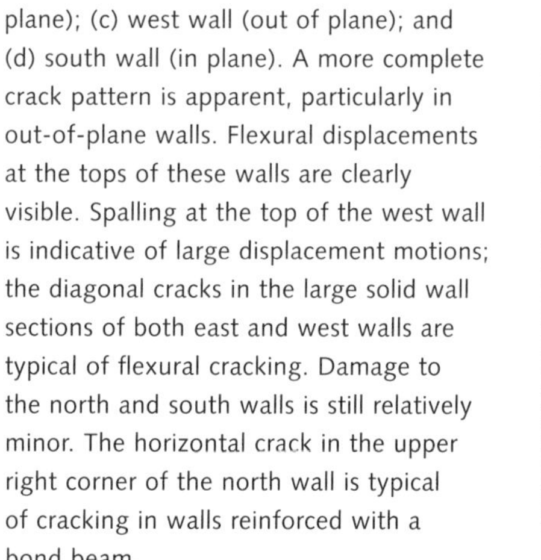

Figure 3.12
Model 6, test level IV, showing (a) east wall (out of plane); (b) north wall (in plane); (c) west wall (out of plane); and (d) south wall (in plane). A more complete crack pattern is apparent, particularly in out-of-plane walls. Flexural displacements at the tops of these walls are clearly visible. Spalling at the top of the west wall is indicative of large displacement motions; the diagonal cracks in the large solid wall sections of both east and west walls are typical of flexural cracking. Damage to the north and south walls is still relatively minor. The horizontal crack in the upper right corner of the north wall is typical of cracking in walls reinforced with a bond beam.

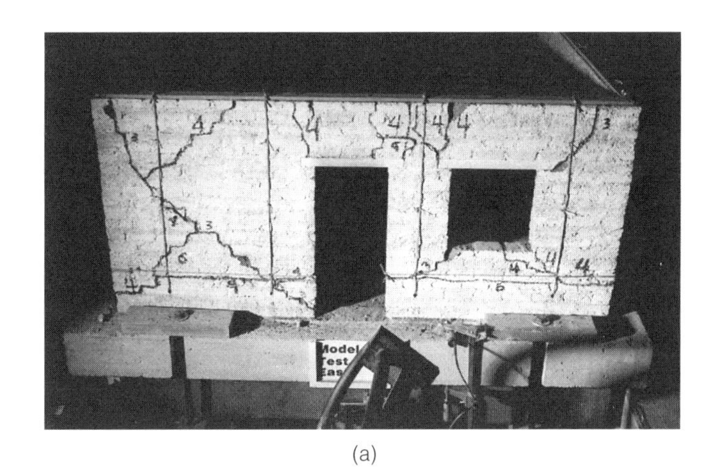

(a)

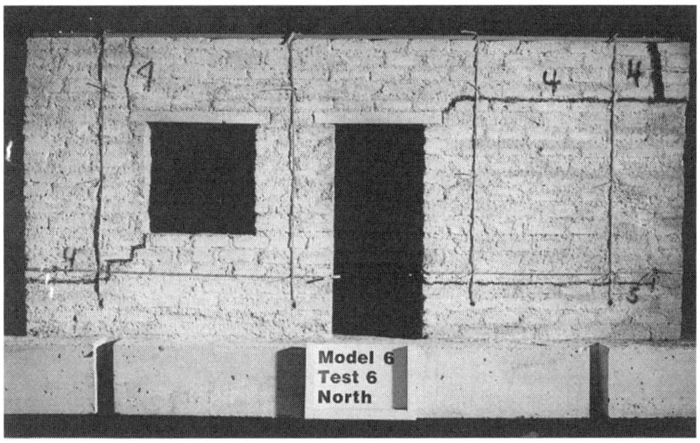

(b)

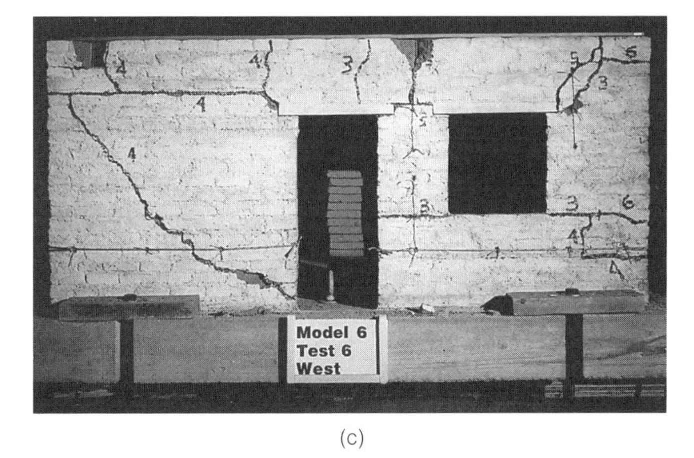

(c)

(d)

Figure 3.13
Model 6, test level VI, showing (a) east wall (out of plane); (b) north wall (in plane); (c) west wall (out of plane); and (d) south wall (in plane). During test levels V and VI, most of the dynamic motions were observed in the out-of-plane walls. Deflections at the tops of the walls were large and completely dependent on the stiffness of the bond beam. In the in-plane direction, the cracks were largely developed, but there was little motion or displacement across cracks.

(a)

(b)

(c)

(d)

Figure 3.14
Model 6, test level VII, showing (a) east wall (out of plane); (b) north wall (in plane); (c) west wall (out of plane); and (d) south wall (in plane). Spalling at the top of the east and west walls is indicative of large out-of-plane displacements. Some additional minor cracks developed during this test. In-plane displacements of north and south walls became more substantial, but little new cracking developed. Some small permanent displacements occurred at the end of this test.

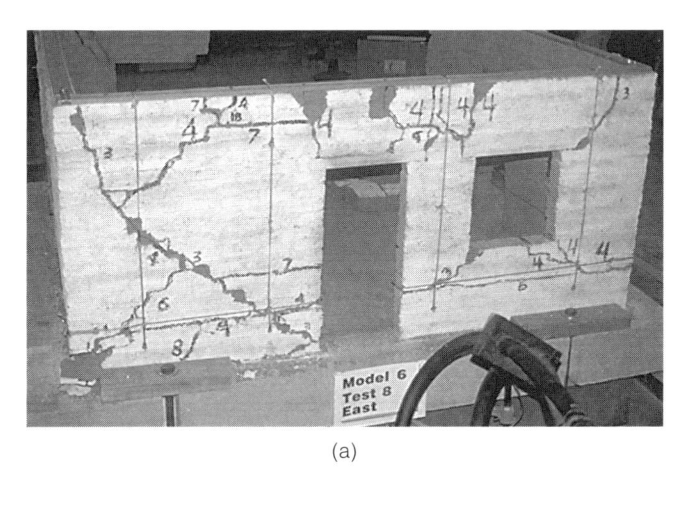

(a)

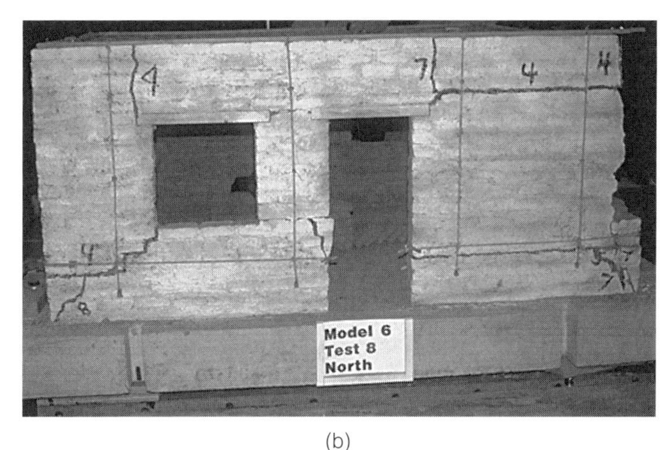

(b)

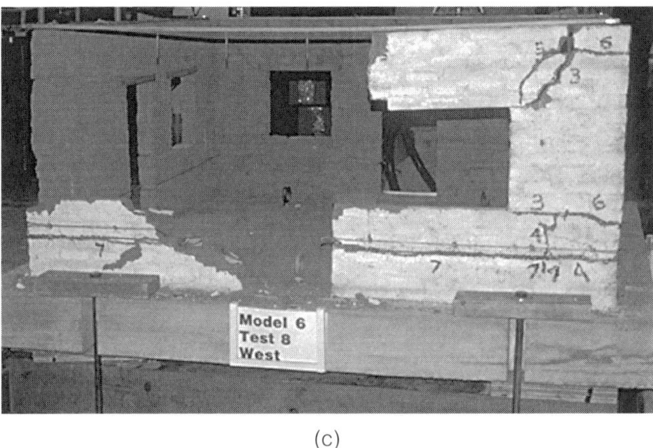

(c)

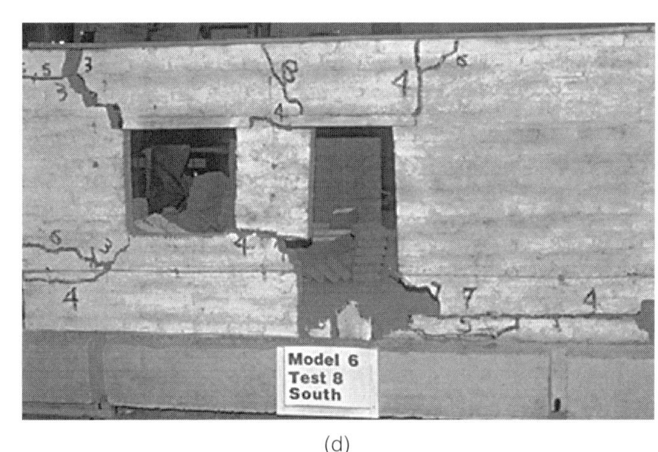

(d)

Figure 3.15
Model 6, test level VIII, showing (a) east wall (out of plane); (b) north wall (in plane); (c) west wall (out of plane); and (d) south wall (in plane). Dynamic motions of the model were substantial. During test levels VI and VII, stability of the out-of-plane walls was dependent on the bond beam. During test level VII, out-of-plane stability became dependent on both the bond beam and the vertical straps. Collapse of the west wall by overturning started below the horizontal crack that formed during test level IV at the north end of the west wall. The north and south walls were also subjected to substantial movement, resulting in significant permanent displacements. Relative displacements of cracked wall sections were less severe in the north wall, which had vertical straps, than in the south wall, which had only local through-ties.

In test level VIII, the lightly retrofitted west wall collapsed (fig. 3.15). Local crack ties had been applied at several locations but were not effective in preventing out-of-plane collapse.

A portion of the in-plane south wall collapsed (fig. 3.16) during test level IX. Nevertheless, the two sections of the north wall that were retrofitted with vertical straps continued to perform well.

Most portions of the north and east walls performed well through test level X (fig. 3.17). The walls did not collapse nor was the building in the unstable condition noted for model 4 after test level X. The pier between the door and window was dislodged during the test and was suspended by the vertical straps. There were insufficient displacement controls on this pier to prevent it from being dislodged.

Summary of Test Results for Models 4–6

Model 4 performed well throughout the testing sequence despite its near collapse during test level X. The performance of models 1 and 4 were nearly the same, even though model 1 did collapse during test level X. The horizontal straps clearly had a positive effect on providing stability for both the in-plane and out-of-plane walls of these models.

The unretrofitted model 5 collapsed early in the test sequence, as expected. The model was not able to sustain motions much

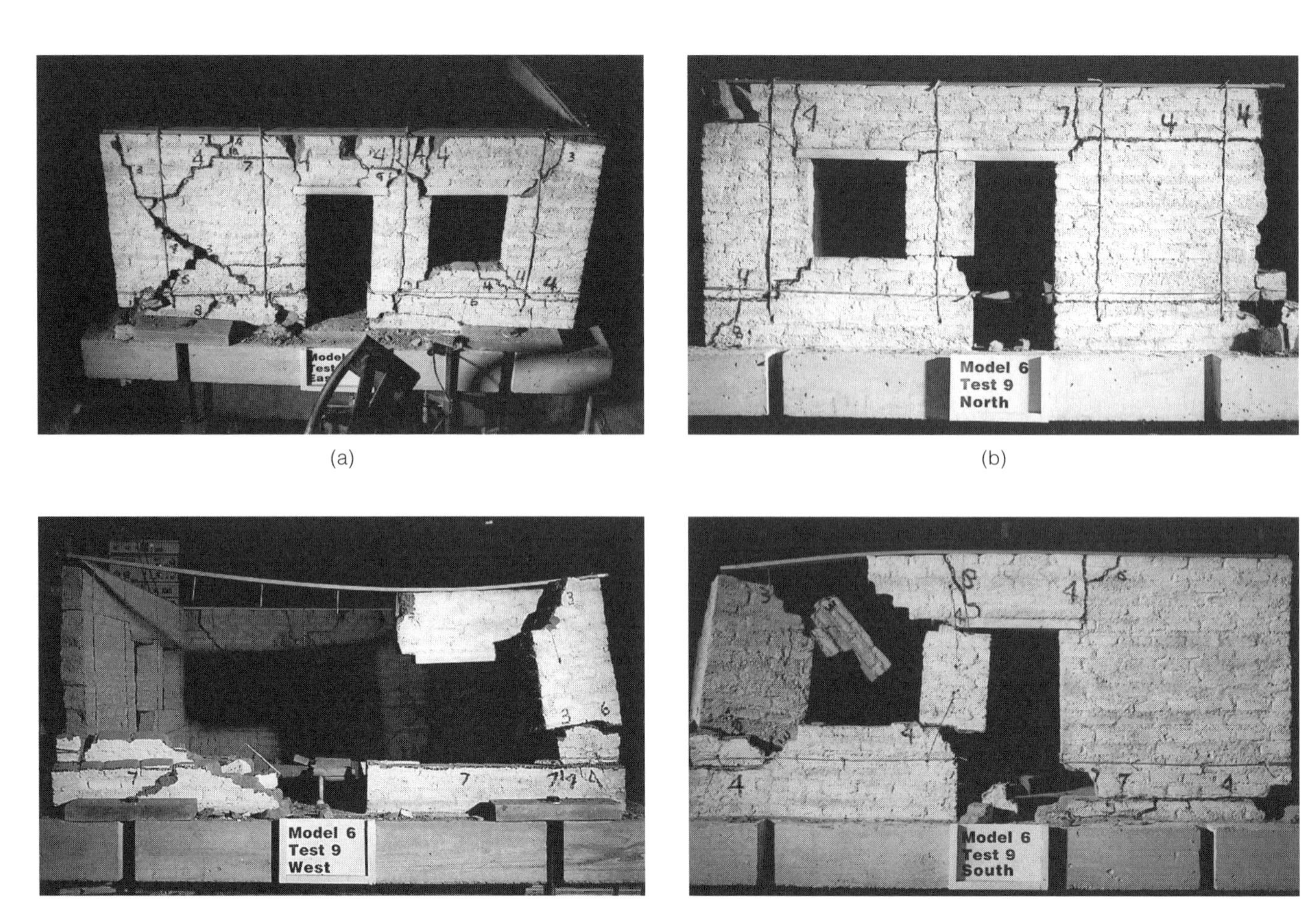

(a) (b) (c) (d)

Figure 3.16
Model 6, test level IX, showing (a) east wall (out of plane); (b) north wall (in plane); (c) west wall (out of plane), collapsed at test level VIII; and (d) south wall (in plane). The test sequence continued, despite the collapse of the west wall, which undoubtedly compromised the structural performance of the south wall. Nevertheless, the continuous successful performance of the north and east walls demonstrated the robustness of the retrofit system used on these two walls. The east wall again showed large displacements, but the dynamics of the wall was also largely affected by displacements of the in-plane walls. Overall displacements in the north wall were smaller than those in the south wall. Local ties have some overall stabilizing influence but seem to be relatively poor as displacement controls.

greater than those that caused the initial cracking; it survived only one test beyond the fully developed cracked state.

Model 6 (and all other retrofitted models) behaved much better than the control model. The walls with vertical straps (north and east walls) behaved considerably better than the south and west walls, which had only local ties. The south and west walls collapsed during test level VIII; the bond beam and the local crack ties improved the performance of the walls slightly but did not provide sufficient stability to prevent collapse. The north and east walls behaved very well through test level IX, but the east wall suffered partial collapse during test level X. The behavior of the north and east walls may have been negatively affected by the earlier collapse of the south and west walls because they would have provided some support, particularly at the south end of the east (out-of-plane) wall.

The behavior of model 2 (S_L = 7.5) and model 6 (S_L = 11) was similar, although the performance of model 2 was clearly more robust due to the additional stability provided by the thicker walls. The addition of vertical straps improved the stability of the north and east walls compared to that of the south and west walls, where only local crossties were used in conjunction with the bond beam.

The most significant difference between the dynamic behavior of model 2 (S_L = 7.5) and model 6 (S_L = 11) was observed in the post-

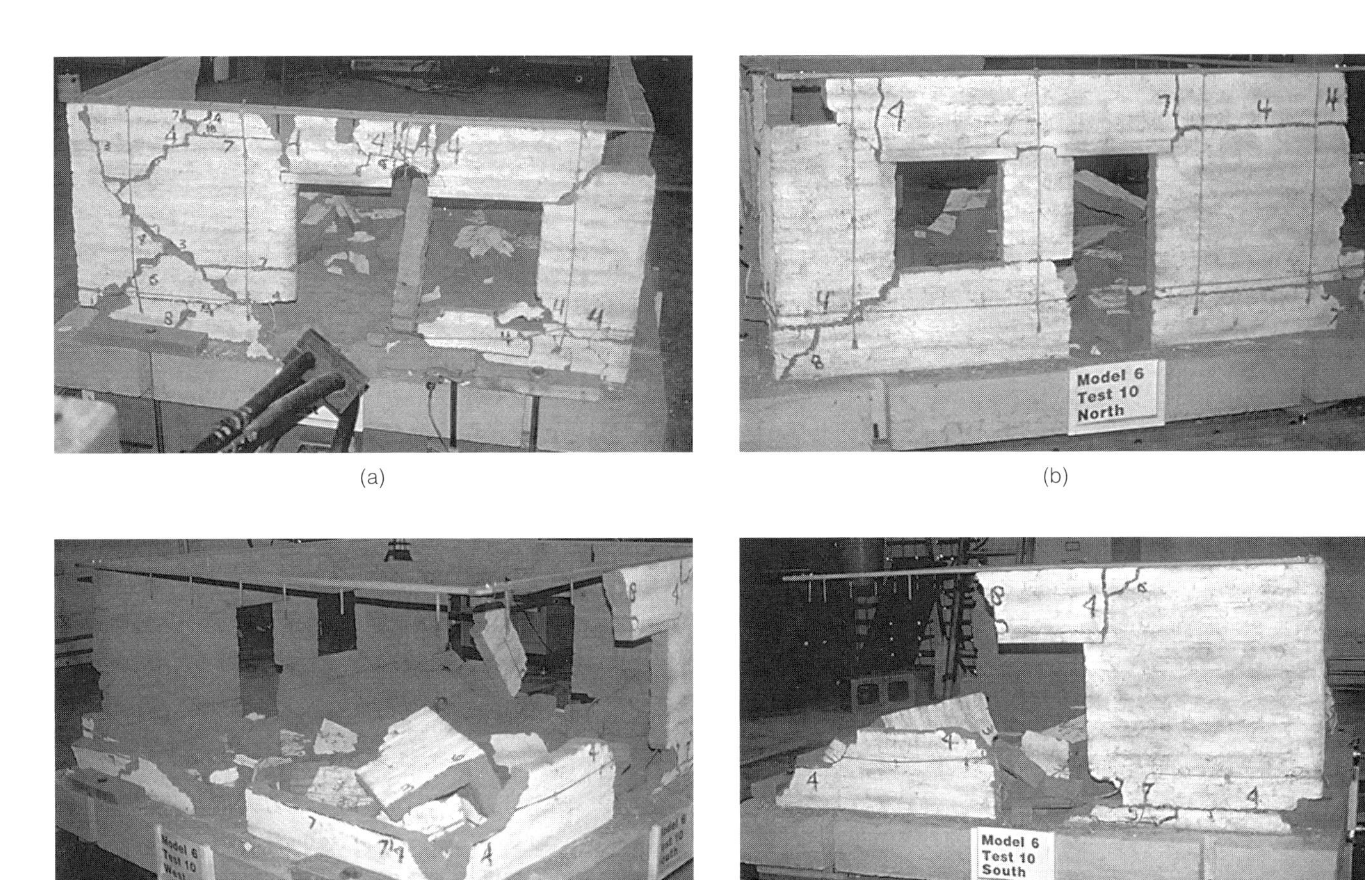

Figure 3.17
Model 6, test level X, showing (a) east wall (out of plane); (b) north wall (in plane); (c) west wall (out of plane); and (d) south wall (in plane). Model approached complete collapse. The north wall behaved well, the bond beam and vertical straps being quite effective. On the east wall, the vertical strap on the central pier did not provide enough stability to prevent collapse. The remainder of the east wall degraded severely, showing that this type of retrofit system for walls with $S_L = 11$ is not as robust as on walls with $S_L = 7.5$ (model 2), which withstood several repetitions at test level X.

elastic dynamic behavior before collapse occurred. After cracks developed, but before stability became a problem (during test levels V–VII), the out-of-plane walls of model 6 showed considerable amplification of the dynamic motion. The frequency of the out-of-plane motion was attributed primarily to the stiffness of the bond beam and the tributary mass of the adobe wall.

In evaluating the effect of wall thickness on the seismic behavior of the retrofitted models, the following conclusions were drawn from these tests:

1. The effect of the installed retrofit systems was of primary importance compared to the overall effect of an increase in wall thickness.
2. The difference in the performance between thin-walled and thick-walled adobe buildings should not be discounted, however. There were some obvious differences in the dynamic behavior of the thin-walled ($S_L = 11$) models compared to that of the thick-walled ($S_L = 5$) models, and the latter required a much lighter intervention than the former.
3. The post-elastic behavior of the thin, out-of-plane walls ($S_L = 11$) was controlled by the dynamic characteristics of

the wood bond beam. The restraint at the base of the wall, which controlled the rocking of wall sections, was negligible compared to the restraint provided by the bond beam. Therefore, the dynamic motion at the top of the wall could be closely estimated by using the tributary mass and the stiffness of the bond beam.

4. In contrast, the out-of-plane behavior of the walls of model 2 ($S_L = 7.5$) did not demonstrate the same dynamic amplification as that of model 6 ($S_L = 11$). The dynamic motion of the out-of-plane walls of model 2 was controlled primarily by the rotational restraint against rocking at the base of the wall.

Chapter 4

Description of Tests for Model 7

The models and tests discussed in the previous chapters were designed primarily to study the behavior of adjoining adobe walls during and following simulated seismic excitation. Model 7, however, was the first model tested in this research project that resembled a real building—with an attic and a roof—that could be used to study the global response of an essentially complete structure (fig. 4.1).

Layout of Model 7

The design of model 7 was based on the tapanco-style adobe architecture commonly seen in the southwestern United States. These historic structures were characterized by walls that extended 61–91 cm (2–3 ft.) above an attic floor and had a pitched roof (8:12) that covered the attic and gable-end walls (S_L = 5). The load-bearing north and south walls each had a door and a window, the west gable-end wall had a small attic window, and the east wall was windowless (figs. 4.1 and 4.2).

Figure 4.1
Model 7 before testing.

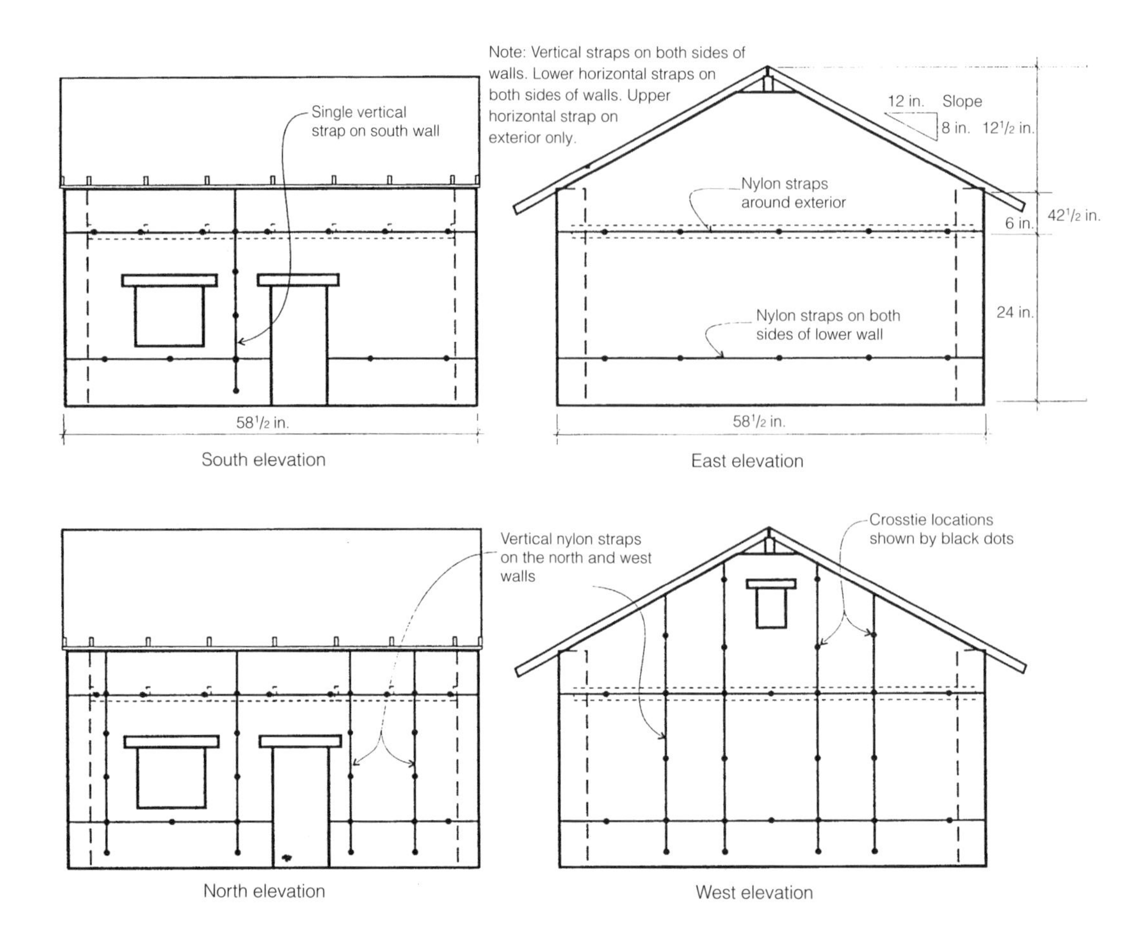

Figure 4.2
Elevations showing retrofit measures applied to model 7. The south wall has only one vertical strap, and the east wall has none, whereas the north and west walls have four vertical straps each.

Retrofit Measures

The retrofit measures used with model 7 were based on the more successful measures tested in models 1–6, with the addition of partial diaphragm systems on the attic floor and roof. The retrofitting system consisted of horizontal and vertical straps applied to the walls, and partial wood diaphragms anchored to the attic floor and roof. The remainder of the retrofit system consisted of connection details. Details of the system are shown in figures 4.3–4.5.

A combination of vertical and horizontal straps was applied to all the walls. As in previous model tests, the retrofit strategy for the south and west walls was slightly different from the strategy for the east and north walls. Two horizontal straps were placed on each of the four walls. The upper horizontal strap was located at the attic-floor level, and the lower horizontal strap was located just below the windowsill. The strap at the attic-floor level was attached to the attic-floor diaphragm (fig. 4.3). The lower horizontal strap was applied on both the exterior and interior surfaces of the walls, and through-wall crossties were used to connect interior and exterior straps.

No vertical straps were applied to the east wall. The south wall had one vertical strap at the center of the pier between the door and the window. The north and west walls each had four vertical straps that were wrapped over the tops of the walls and through holes drilled at

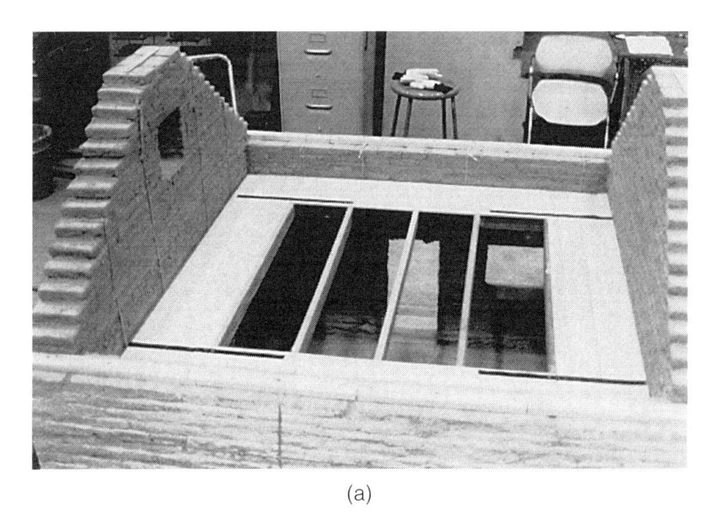

(a)

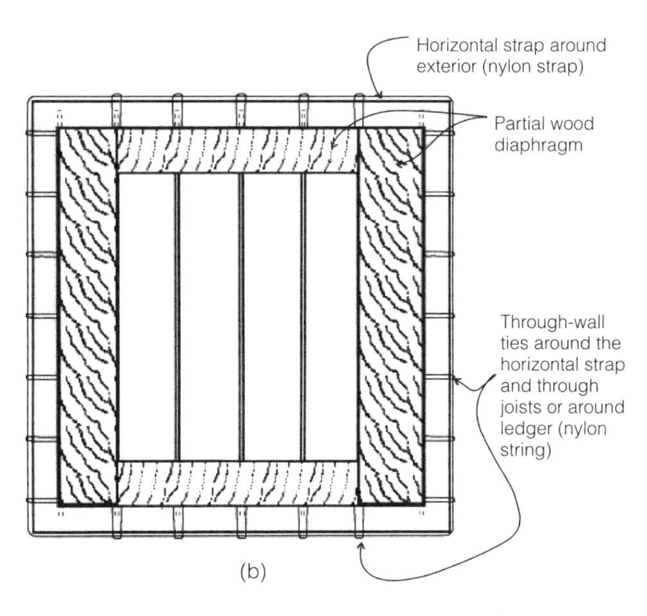

(b)

Figure 4.3
Model 7: (a) photograph of attic-floor framing and partial diaphragm, made of wood; and (b) attic plan, showing retrofit measures applied at floor level.

(a)

(b)

(c)

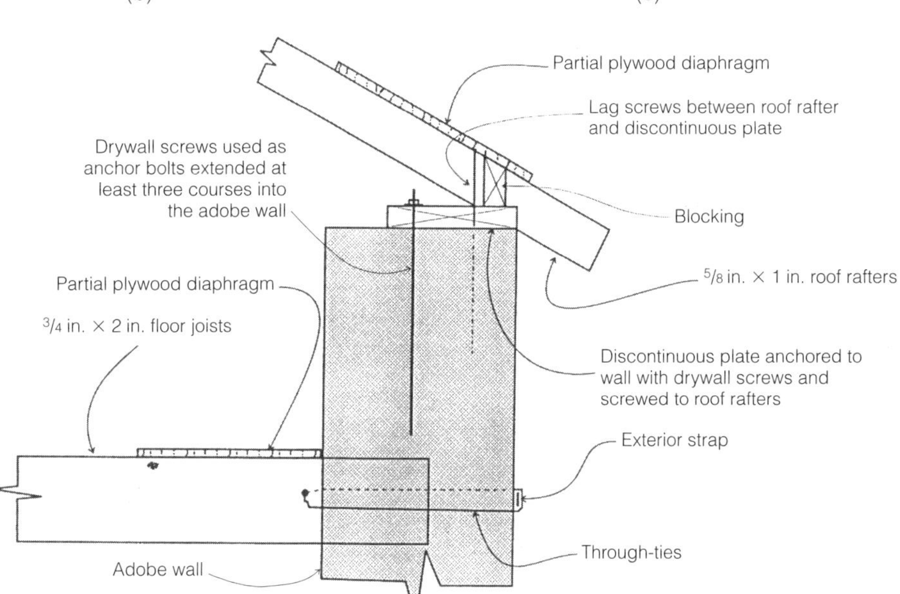

(d)

Figure 4.4
Details of model 7: (a) inside of attic floor, showing through-ties connecting joist and wall; (b) roof line of load-bearing wall; (c) interior corner of partial wood diaphragm; and (d) section drawing of wall, roof, and attic floor, showing retrofit measures.

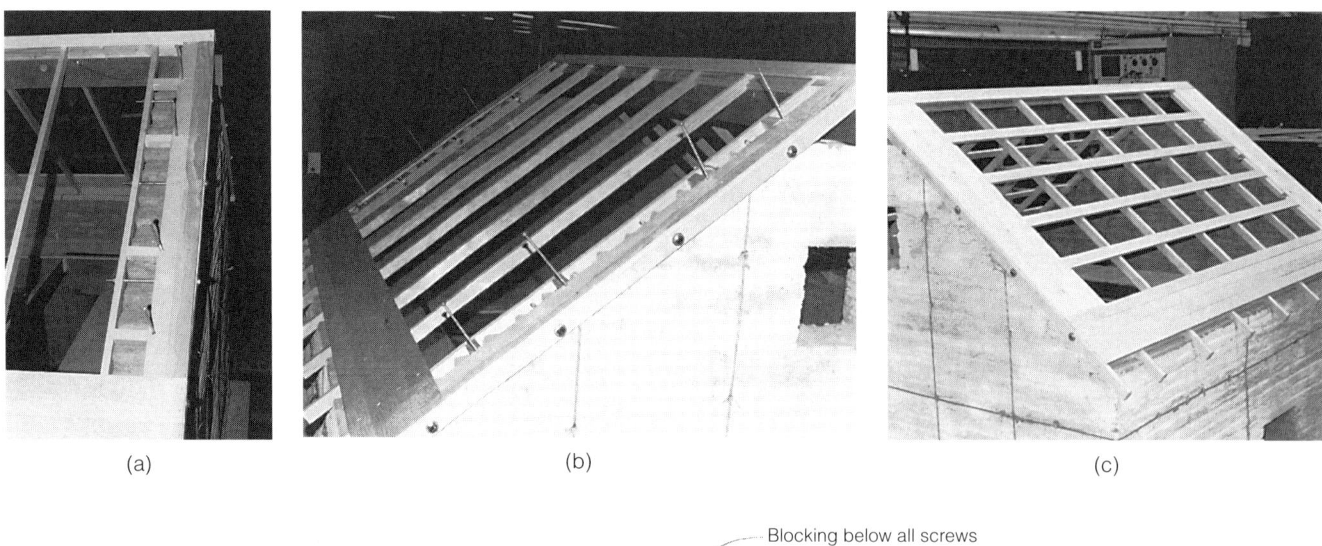

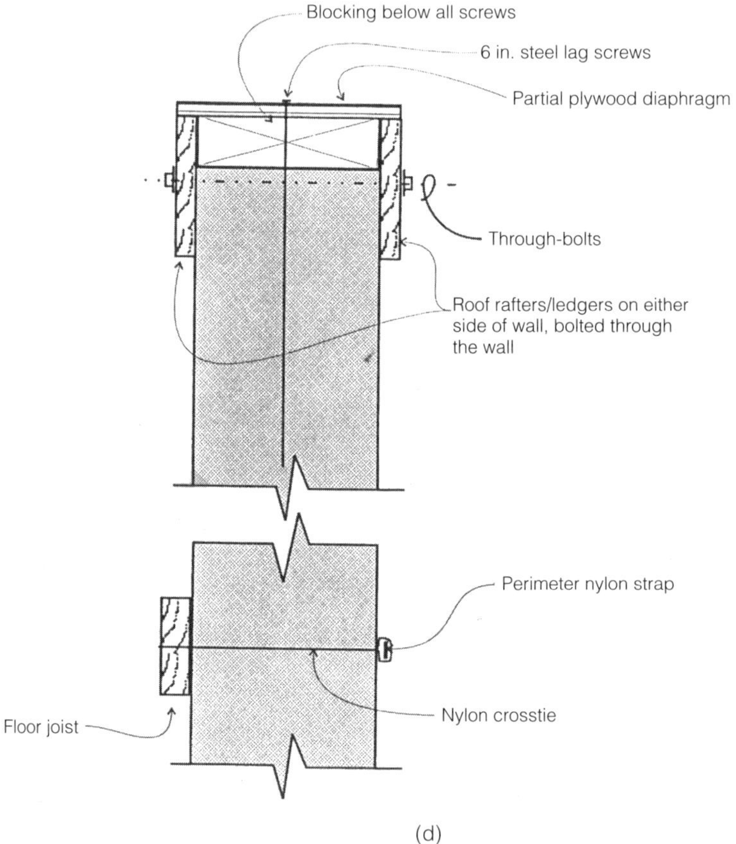

Figure 4.5
Model 7 connections at non-load-bearing walls: (a) detail at east gable-end wall, showing anchor bolts and one-half of partial diaphragm exposed; (b) view of south roof system and west wall details; (c) completed roof system; and (d) section drawing of wall. The nylon crosstie connects the horizontal perimeter strap to the floor joist through holes in the wall and joist.

their bases. Small-diameter straps were used as crossties in a manner similar to those used with the lower horizontal straps.

Partial wood diaphragms were added to the attic floor and roof. The width of the attic-floor diaphragm was approximately 20 cm (8 in.) and was equivalent to the spacing between the floor joists. The width of the partial roof diaphragm was approximately 15 cm (6 in.).

The bearing plates on the tops of the load-bearing walls (fig. 4.4) were segmented so as not to act as stiffening flexural elements. The plates were cut into four segments on the tops of the north and south walls and were anchored to the walls with 7.6 cm (3 in.) long, coarse-thread (gypsum wall board) screws.

The attic-floor joists were anchored to the walls with small-diameter through-wall ties that were threaded through small holes in the center of the attic-floor joist, as shown in figure 4.4a. The cord was passed through the adobe wall on both sides of the joist and was attached to the horizontal strap on the exterior face of the wall. The roof rafters were anchored with screws to the bearing plates, and blocking was placed between each of the roof rafters (fig. 4.4b).

On the non-load-bearing walls, the roof rafters were placed directly on both sides of the wall and tied together with bolts through the wall. The partial roof diaphragm was attached to the tops of the roof rafters. Coarse-thread screws 15 cm (6 in.) long extended through the roof diaphragm and blocking into the wall (see fig. 4.5). The purpose of these details was to anchor the tops of the gable-end walls to the roof system. These connections worked well and did not fail during the tests.

Summary of Test Results for Model 7

Overall, the performance of model 7 and the behavior of the retrofit measures were very satisfactory (figs. 4.6–4.12). A review of the videotape that recorded wall motions during the tests indicated that substantial sections of the model would have collapsed during test level VI or VII.

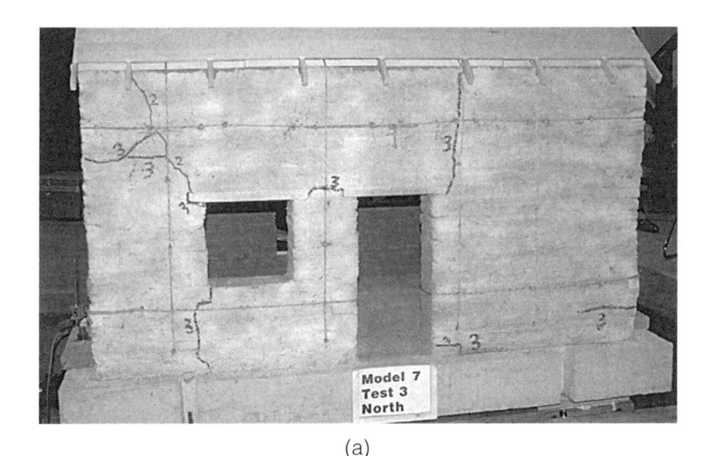

(a)

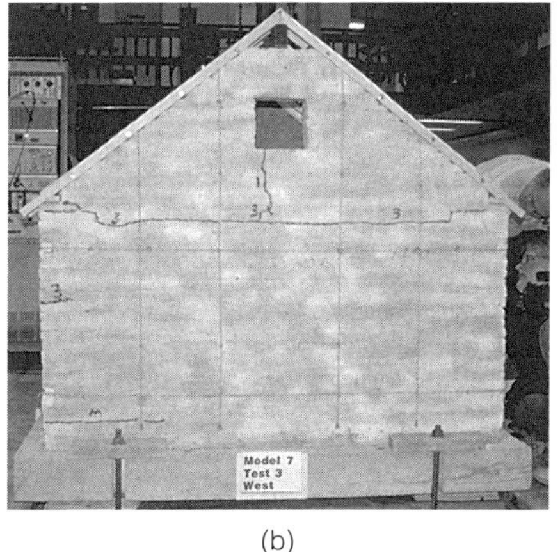

(b)

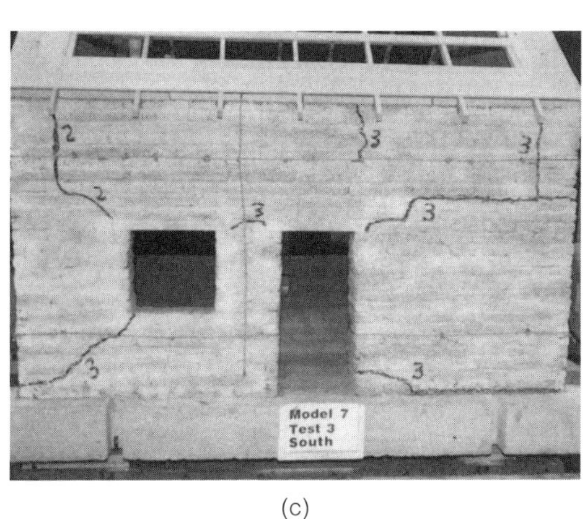

(c)

(d)

Figure 4.6
Model 7, test level III, showing (a) north wall (in plane); (b) west wall (out of plane); (c) south wall (in plane); and (d) east wall (out of plane). Initial cracking is visible. Shear cracks developed in the north and south walls at the corners of openings and at the base of the large wall panel to the right of the doorway. In the east and west walls, a single horizontal crack developed in each wall. In the east wall, the crack was just below the attic-floor line. In the west wall, the crack was just above the attic-floor line.

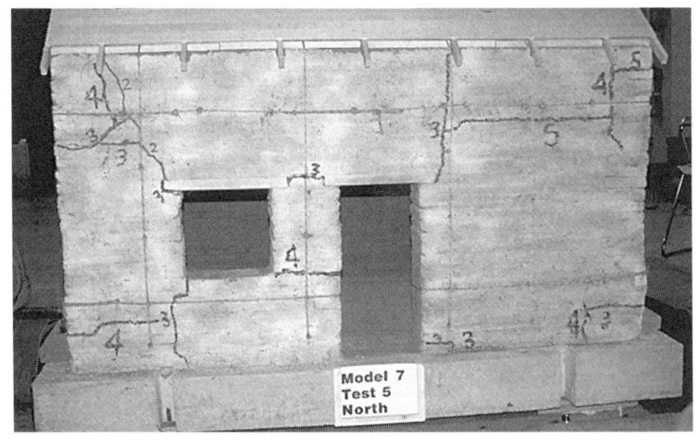

(a)

(b)

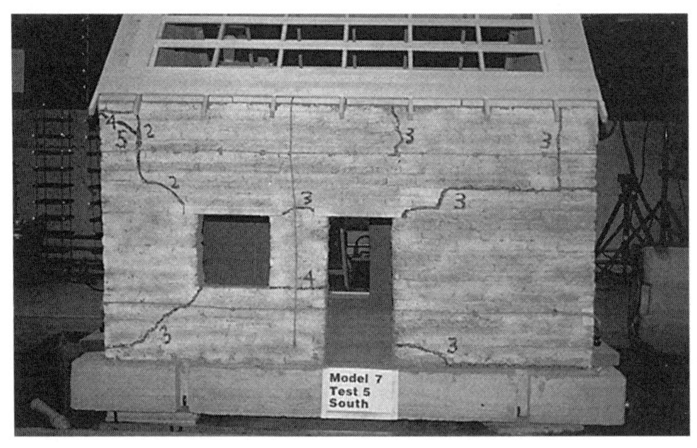

(c)

(d)

Figure 4.7
Model 7, test level V, showing (a) north wall (in plane); (b) west wall (out of plane); (c) south wall (in plane); and (d) east wall (out of plane). Most new damage was in the east and west walls. The upper section of the walls continued to rock, and vertical cracks developed near the center of both the east and west walls. An additional horizontal crack developed in the east wall above the attic-floor line. The north and south wall damage remained relatively constant, although a new horizontal crack developed in the north wall in the large panel at the attic-floor level.

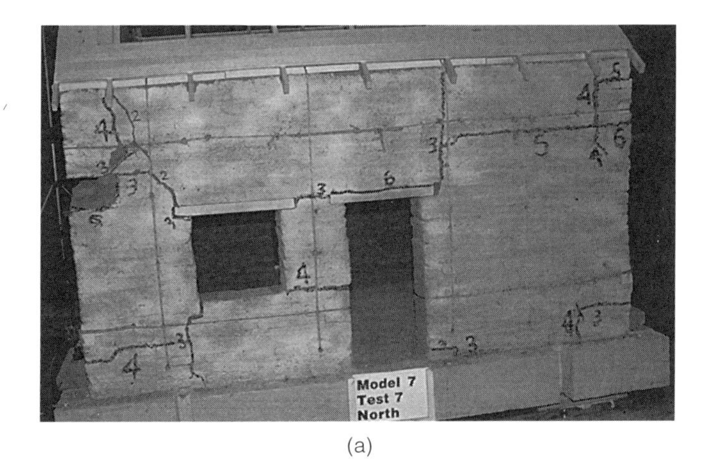

(a)

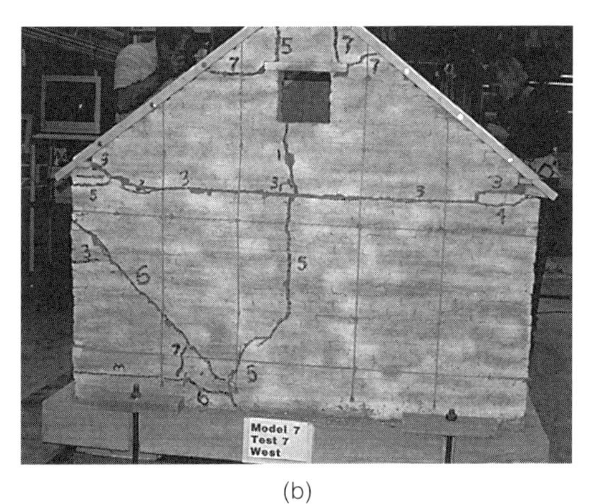

(b)

(c)

(d)

Figure 4.8

Model 7, test level VII, showing (a) north wall (in plane); (b) west wall (out of plane); (c) south wall (in plane); and (d) east wall (out of plane). Major crack patterns were nearly complete after test level V. During test levels VI and VII, some cracks were extended, but most remained unchanged. The principal exceptions were a diagonal crack in the west wall near the north end during test level VI and a similar crack in the east wall during test level VII. Only a few new cracks developed in the north and south walls during these tests. During test level VII, a section in the upper east corner of the north wall dislodged, and a permanent displacement occurred at the lower east corner of the window on the north wall.

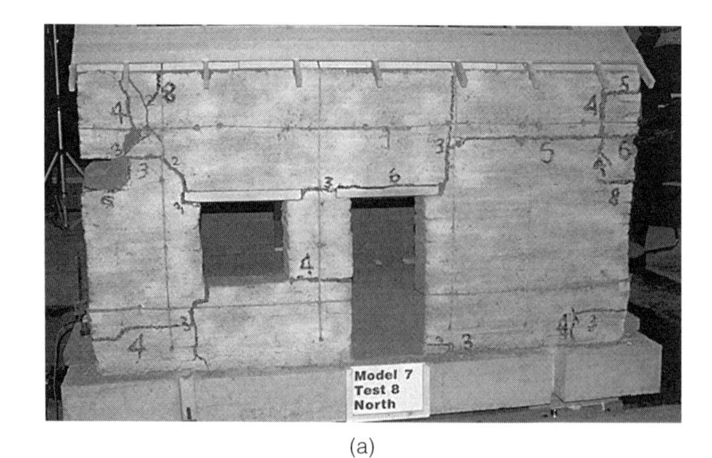

(a)

(b)

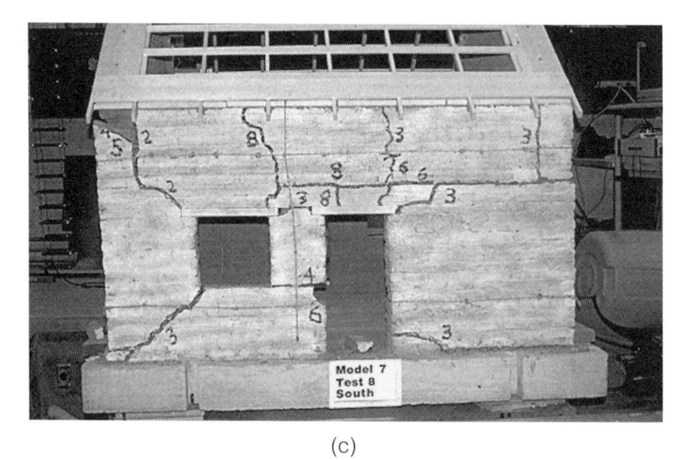

(c)

(d)

Figure 4.9

Model 7, test level VIII, showing (a) north wall (in plane); (b) west wall (out of plane); (c) south wall (in plane); and (d) east wall (out of plane). Additional cracks developed in the east and west walls, notably a horizontal crack in the lower portion of the west wall and a vertical crack in the right lower portion of the east wall, which was a continuation of an existing vertical crack in the upper gable portion of the wall. Additional cracks developed in the north and south walls. Of further significance were the permanent offsets of 0.64–1.27 cm (0.25–0.5 in.) at some locations.

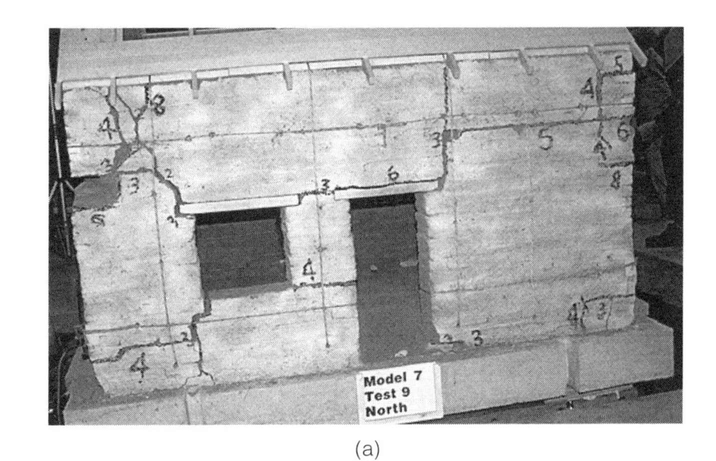

(a)

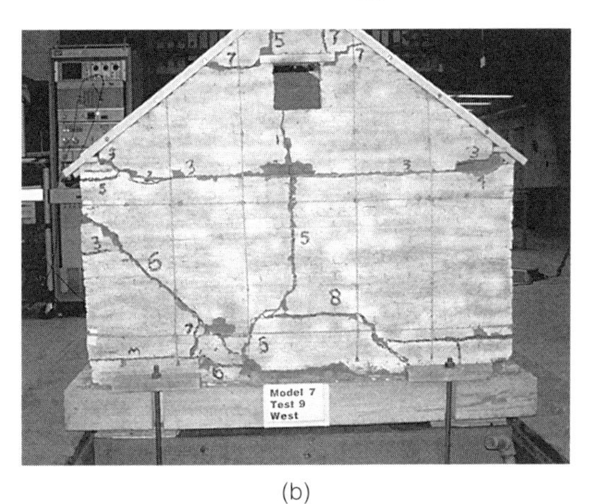

(b)

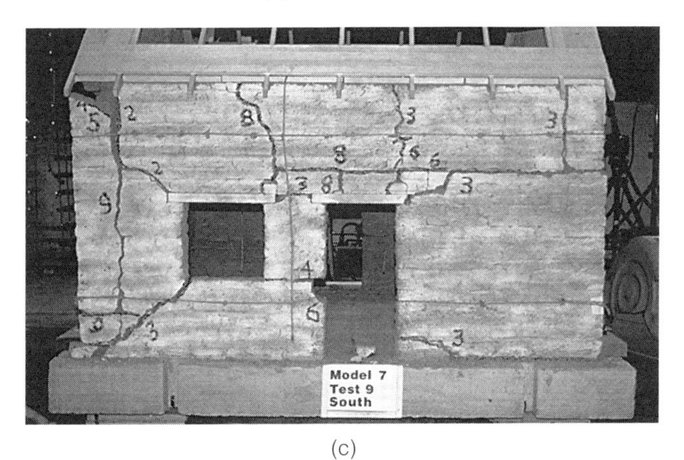

(c)

(d)

Figure 4.10
Model 7, test level IX, showing (a) north wall (in plane); (b) west wall (out of plane); (c) south wall (in plane); and (d) east wall (out of plane). Only a few new cracks developed in the model during the test, and wall motions were confined mainly to sliding and rocking of the cracked wall sections. In the east and west walls, some permanent offsets occurred at the horizontal cracks. The permanent offsets observed in the north and south walls continued to increase.

Figure 4.11
Model 7, test level X(1), the first of two tests at this level, showing (a) north wall (in plane); (b) west wall (out of plane); (c) south wall (in plane); and (d) east wall (out of plane). The dynamic stability of the structural system is clearly in question here. Horizontal offset at the attic-floor level in the east wall is most visible on the east side of the south wall elevation. The condition of the north and south walls continued to degrade. The section of the wall to the left of each window had offsets of approximately 1.88 cm (0.75 in.). The offsets at the upper west side of the south wall were largest. Both north and south walls had offsets of approximately 3.8 cm (1.5 in.) along the diagonal crack at the lower left corners of the windows.

Instead, the model performed well through test level X. Only a lightly retrofitted section of the south wall collapsed during the first repetition of test level X. Table 4.1 is a summary of the model's performance. The important aspects of the performance of model 7 are as follows:

1. The model behaved very well and generally as expected based on test results of the previous six models. The retrofit system used on this model was clearly a success.
2. The cracking pattern was generally consistent with that observed in previous models, but some elements were not expected. The inclusion of vertical and horizontal straps, with crossties at regular intervals, allowed the retrofit measures to behave well even when cracks did not occur where they might have been expected.
3. The partial roof diaphragm prevented out-of-plane collapse of the gable-end walls. Large displacements occurred at the tops of these walls due to the flexibility of the roof system. The roof diaphragm was quite flexible due to the discontinuity of the diaphragm at the ridge beam.

Figure 4.12
Model 7, test level X(2), the second of two tests at this level, showing (a) north wall (in plane); (b) west wall (out of plane); (c) south wall (in plane); and (d) east wall (out of plane). Lack of restraint in the southeast corner allowed the unstable wall section to collapse at the end of this test. The retrofit measure on the north and west walls, where vertical straps were used, provided sufficient restraint on the cracked wall sections to ensure overall structural stability.

4. The partial diaphragm at the attic level was considerably stiffer than the roof diaphragm. The through-wall connections performed well. Horizontal cracks developed in the two gable-end walls, and the base was able to rock and slide at these locations.
5. Permanent displacements of 2.5–5 cm (1–2 in.) occurred at these horizontal cracks in the walls following test levels VIII–X. The retrofit system was sufficiently effective to prevent collapse of these walls but not to prevent displacement.
6. The lower horizontal straps worked effectively to prevent the deterioration of the piers under the windows. Diagonal cracks extended from the lower outer corners of the windows downward to the northeast and southwest corners, but the straps prevented substantial widening of these cracks.

Table 4.1
Model 7: Summary of performance

Test level	Maximum EPGA (g)	Maximum displacement cm	in.	Performance summary
I	0.12	2.54	1.00	No damage
II	0.18	5.08	2.00	Minor cracks initiated on north, south, and west walls
III	0.23	7.62	3.00	Further development of cracks; first crack developed in east wall—a horizontal crack at the attic-floor line
IV	0.28	10.16	4.00	Relatively minor additional cracks, most in the in-plane (north and south) walls
V	0.32	12.70	5.00	Substantial additional cracking; vertical cracks developing in both gable-end walls; minor additional cracks in the in-plane (north and south) walls
VI	0.40	15.88	6.25	Crack pattern nearing full development; negligible crack offsets
VII	0.44	19.05	7.50	Measurable but minor horizontal offsets of horizontal cracks at attic-floor line in gable-end walls
VIII	0.48	25.40	10.00	Offsets in the in-plane and out-of-plane walls
IX	0.54	31.75	12.50	Worsening of cracks and spalling of adobe in corners at floor line and at tops of walls
X(1)	0.58	38.10	15.00	Substantial offsets, particularly in east and south walls, which did not have vertical straps; offsets up to 2.54 cm (1 in.) (model dimensions)
X(2)	0.58	38.10	15.00	Collapse of lower section of south wall

Chapter 5

Description of Tests for Models 8 and 9

The tests on models 8 and 9 were designed to continue refinement of retrofit strategies using the more complex, tapanco-style buildings. The tests on model 8 were designed to determine whether the strategy used on model 7 would be successful on a building with thinner walls. In addition, the south and west walls of the models would be retrofitted with vertical center-core rods instead of vertical straps. Model 9 was unretrofitted to demonstrate how and at what level of shaking an unretrofitted adobe building would fail under the same test conditions used on models 7 and 8.

Layout of Models 8 and 9

The layout of models 8 and 9 (S_L = 7.5) was the same as that of model 7 (S_L = 5) (except that an attic window was added to the east wall) and was typical of tapanco-style adobe construction. The plans and elevations for these models are shown in figure 5.1. The walls were one wythe wide and the bricks were laid in a running bond pattern.

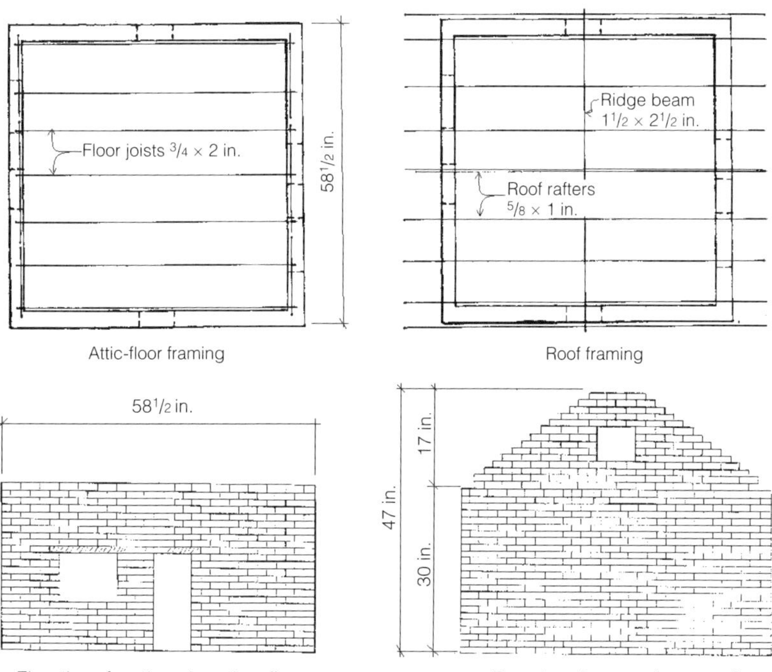

Figure 5.1
Plan and elevations for models 8 and 9.

Construction details are shown in figures 5.2–5.5. In model 8, the floor framing and ridge beam were installed before the roof rafters were in place (fig. 5.2). Figure 5.3 shows model 9 with the complete roof system, which represents the basic structural system to which the retrofit measures were added in model 8. For construction purposes on the small-scale models, much of the retrofit systems for models 7 and 8 needed to be installed before the roof was constructed. The roof sheathing was widely spaced to ensure that the roof system did not significantly affect the out-of-plane motions of the gable-end walls.

Roof-end rafters were placed on both sides of the gable-end walls so the walls were partially bearing against them (figs. 5.4 and 5.5). Through-bolts were used only on model 8. Even without these through-bolts, the joists did provide some lateral restraint to the tops of the

Figure 5.2
Model 8 before roof framing was added, with ridge beam in place.

Figure 5.3
Model 9 after roof framing was installed.

Figure 5.4
Detail of model 8, with through-bolts tying roof rafters together on both sides of the gable-end walls. This retrofit measure was used only on this model.

Figure 5.5
Blocking between roof rafters screwed to sill plate, used on models 8 and 9.

gable-end walls in model 9. Because of this restraint, it was important to provide only minimal horizontal stiffness in the roof system.

The roof rafters were placed on a discontinuous wood top plate that rested on top of the wall. The plate was segmented so as not to act as a bond beam, which would have added strength and stiffness. In typical historic buildings, the top plate is not continuous along the length of the building. Blocking was placed between the rafters at the bearing-wall top plate and at the ridge beam.

Model 8 Retrofit Measures

Model 8 was retrofitted with measures similar to those used on model 7. As in previous models, the retrofit strategy was different on the west and south walls than that of the east and north walls (fig. 5.6). The retrofit system for model 8 consisted primarily of the following:

1. Partial wood diaphragms on the attic-floor and roof framing, with an exterior, horizontal nylon strap attached to the attic-floor framing at the floor level.
2. Lower horizontal and vertical straps on each of the north and east walls.
3. Vertical fiberglass center-core rods placed in holes drilled in the south and west walls, rods epoxy-grouted in place, and a lower horizontal strap placed in a center-core hole of the south wall.

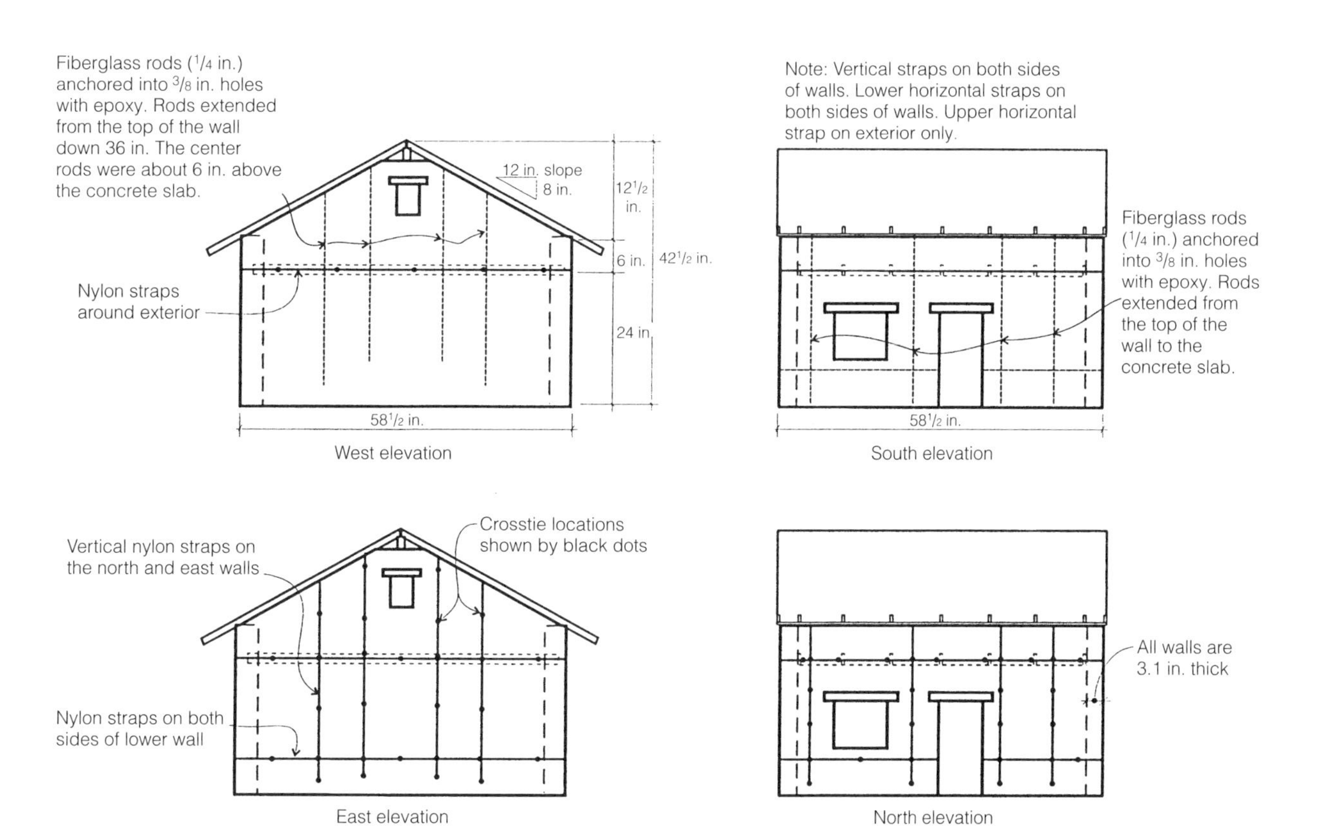

Figure 5.6
Elevations showing retrofit measures applied to model 8.

Figure 5.7
Model 8 before roof framing was completed, with ridge beam in place.

Figure 5.8
Partial diaphragm at the southeast corner of model 8. A nylon strap connected adjacent sections of the wood diaphragm.

Details of many aspects of the retrofit system and its installation for model 8 are shown in figures 5.7–5.15. The connection details were nearly the same as those for model 7 (figs. 4.3–4.5).

Model 8 before the roof diaphragm was added is shown in figure 5.7. Figure 5.8 shows the attic-floor diaphragm at one of the corners. The partial attic-floor diaphragm in model 8 was not as wide as that used in model 7. In model 8, the floor diaphragm was 11.4 cm (4.5 in.) wide; in model 7 it was 19 cm (7.5 in.). A nylon strap was added to ensure transfer of the horizontal loading at the corner of the diaphragm. As demonstrated by the tests, this diaphragm size provided adequate stiffness to control the wall loads and displacements.

The roof framing and partial roof diaphragm are shown in figure 5.9. The roof diaphragm was only 11.4 cm (4.5 in.) wide over the gable ends and 19 cm (7.5 in.) wide over the bearing walls. The discontinuity in the diaphragm at the ridge along the gable-end wall allowed large displacements at the ridge in model 7. To help control these displacements, diagonal sheathing was added at each of the four lower corners of the roof to stiffen the partial diaphragm.

An upper horizontal strap was placed at the attic-floor line and encircled the perimeter of the building (fig. 5.6). This was attached to the floor framing, which was connected to the partial floor diaphragm.

On the north and east walls of model 8, the lower horizontal strap was located on both sides of each section of wall at a level approximately two courses below the bottom of the window. These walls had four vertical straps on each wall, again with nylon straps used as cross-ties, as shown for the east wall in figure 5.6.

The south and west walls were retrofitted using measures designed to minimize damage to the wall surfaces. The vertical elements were fiberglass center-core rods embedded in an epoxy grout at the same intervals as used for the vertical straps on the north and east walls. The only exterior element on the south wall (fig. 5.6) and west wall (fig. 5.10)

Figure 5.9
Model 8 after the final retrofit. Diagonal bracing was added in the corners to stiffen the partial roof diaphragm.

Figure 5.10
West gable-end wall of model 8 after retrofit system was completed. The only sign of the retrofit system on this wall is the horizontal strap at the attic-floor line.

Figure 5.11
Center-core holes being drilled in the west gable-end wall.

Figure 5.12
Epoxy resin grout being injected into model. This measure was used to attain continuity between the center-core rods and the adobe wall.

was a horizontal strap at the attic-floor line. The lower horizontal strap on the south wall was placed inside a horizontally drilled hole. The west wall had no lower horizontal strap at this level because a drill bit of sufficient length was unavailable. Drilling of the holes and injection of the epoxy grout are shown in figures 5.11 and 5.12, respectively. The center-core rods in the west wall terminated just short of the wall base because the drill bit was not long enough to reach the slab level. The outside center-core rods extended to about 5 cm (2 in.) above the slab level, and the inner center-core rods extended to approximately 12.7 cm (5 in.) above the slab. On the south wall, the center-core holes were drilled to, but not into, the slab.

Figure 5.13
Piers on either side of the windows, showing horizontal cuts made to encourage horizontal cracks at these locations.

Horizontal saw cuts were placed on both sides of the piers adjacent to the windows (fig. 5.13). In model 3, saw cuts were used in combination with center-core rods in an attempt to control crack locations. In that test, the saw cuts were successful in controlling cracks at these locations. The test on model 3 also showed the favorable structural performance of the combination of a horizontal crack with vertical center-core elements in controlling the displacement along the cracks.

Small sections of wire mesh were used at a number of corner locations throughout the building to provide a wide bearing area for the nylon straps. Wire mesh is flexible and ductile, and it distributes the load over a wide enough area to prevent the strap from digging into the adobe wall (figs. 5.14 and 5.15).

Figure 5.14
West gable-end wall of model 8, showing wire mesh used for bearing load distribution under vertical straps at the top of the walls.

Figure 5.15
Wire mesh used under upper horizontal straps at the building corners.

Instead of the nylon cord used in model 7, the crossties used in model 8 were commercial nylon cable ties that are easily installed and tightened. Larger cable ties and an automatic tensioning tool could be used in the field. They worked successfully during the tests on model 8. Cable ties were closely spaced above and below the attic floor line to try to control the magnitude of the crack offsets along horizontal cracks that were likely to occur near the floor line.

Test Results for Model 8

The retrofit system for model 8 was designed, in part, to control permanent wall offsets and difficult-to-repair damage. All walls behaved well during the entire test sequence. In the areas where center-core rods were placed, damage was very limited. In other areas, significant offsets occurred and gaps opened between cracked wall sections, but the overall structure was stable during all tests. Although the walls of model 8 ($S_L = 7.5$) were thinner than those of model 7 ($S_L = 5$), the gable-end walls performed very well. There was a somewhat greater displacement control provided by the horizontal straps due to the closer spacing of the crossties and the use of wire mesh. Where mesh was used, straps were prevented from digging into the walls, which would have allowed the straps to loosen and permit larger displacements.

The most surprising result observed was the effectiveness of the epoxy-anchored, center-core rods, which acted as effective reinforcing elements. Visible cracks in walls retrofitted in this way were delayed until much later in the testing sequence than observed in other models (see following sections).

A summary of the important aspects of the results of the test sequence for model 8 is listed in table 5.1. Photographs of model 8 after each test, starting with test level III, are shown in figures 5.16–5.24.

Flexural Test of a Wall Element with Center-Core Reinforcement

After test level X was completed, examination of the west wall was undertaken in an attempt to understand more about the effectiveness of the center-core rods. When the adobe material was broken away from around one of the rods (fig. 5.25), the uneven surface that the epoxy formed at the adobe interface was evident. This uneven surface was indicative of the nature of the bonding of the epoxy resin and the fiberglass rod to the adobe wall, with the result that the composite structure acted effectively as a reinforcing rod. It should be noted that there was a small yet visible crack in the adobe at the time the section of wall was removed from the building. In concrete, reinforcing bars do not become effective in tension or flexure until the concrete has cracked. These cracks can be very small and not even visible to the naked eye. Therefore, even though the section of adobe wall was cracked, it did not affect the strength of this "reinforced adobe beam."

Table 5.1

Model 8: Summary of performance (displacements and accelerations in prototype dimensions)

Test level	Maximum EPGA (g)	Maximum displacement cm	in.	Performance summary
I	0.12	2.54	1.00	No damage
II	0.18	5.08	2.00	Very minor cracks in west gable-end wall below window
III	0.23	7.62	3.00	Minor cracking in east wall near window; no cracks in west wall; crack development in north wall at openings; only one minor crack formed in the south wall
IV	0.28	10.16	4.00	Development of vertical cracks in both out-of-plane walls with some cracking at the base; development of additional cracks, typical at this stage of the testing sequence, in north (in-plane) wall; only damage to the south wall: vertical crack caused by tensile stresses from pullout of the west wall
V	0.32	12.70	5.00	Development of diagonal yield-line cracks in out-of-plane walls; no significant new cracks in the in-plane walls, though significant cracks already present in north wall; no damage in south wall except existing vertical crack near the west wall
VI	0.40	15.88	6.25	No new cracks observed in south wall; some additional horizontal cracks formed in the already badly damaged north wall; development of additional cracks in out-of-plane walls: some horizontal cracks in east wall and cracks around the two main panels reinforced with center-core rods in west wall
VII	0.44	19.05	7.50	Still no new cracks in the nearly undamaged south wall; a few additional cracks formed in the severely cracked north wall although with no significant offsets; minor additional cracks in the east and west out-of-plane walls
VIII	0.48	25.40	10.00	Appearance of some cracks in south wall, however, main panels in the wall with center-core rods still not cracked; offsets of 0.6 cm (0.25 in.) at three locations in north wall; offsets of approximately 0.6 cm (0.25 in.) in east wall (straps); no offsets observed in west wall (center-core rods)
IX(1)	0.54	31.75	12.50	Retrofit system very effective, although offsets beginning to grow larger at most locations; offsets in north and east walls (straps) increased by about 0.3 cm (0.13 in.); offsets in west wall (center-core rods) minimal except at base below center cores; still only minimal damage to south wall except for mid-height block at corner, showing offset of approximately 0.95 cm (0.38 in.)
IX(2)	0.54	31.75	12.50	Main areas of offsets growing in each wall, although building continuing to perform well; development of horizontal crack in north block in west wall (center-core rods) but because of reinforcing effects of center cores, crack remains small
X	0.58	38.10	15.00	Model performing well, with offsets in each wall increasing up to 2.5 cm (1 in.); sections of wall below ends of center-core rods in west wall dislodged

The section of the west wall was removed from the model and then statically loaded at mid-span with lead blocks until failure occurred (fig. 5.26). A videotape made during the tests showed that the adobe material was crushed when the beam failed. The load at failure was approximately 72.6 kg (160 lb.). Based on calculations similar to those used for analyzing concrete, the capacity of this beam was approximately 124 N-m (1100 in.-lb.). The moment induced by the loading was approximately 136 N-m (1200 in.-lb.). These calculations demonstrate an accurate analytical method for determining the strengthening effects of the fiberglass bars embedded in the adobe wall of the model building.

Test Results for Model 9

Model 9 was an unretrofitted, tapanco-style control model similar to model 8. The results of the sequence of tests on model 9 demonstrated

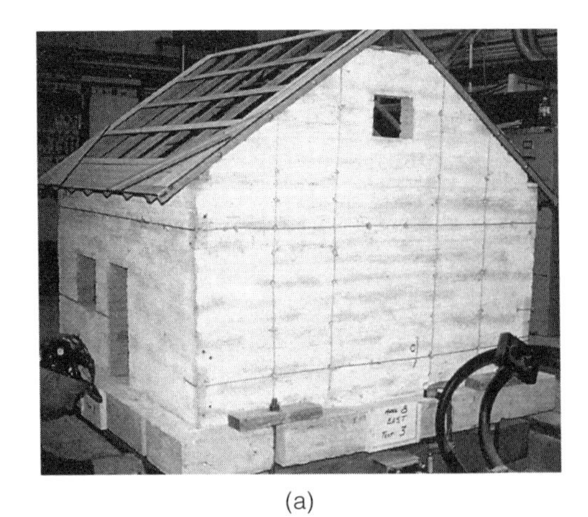

(a)

(b)

(c)

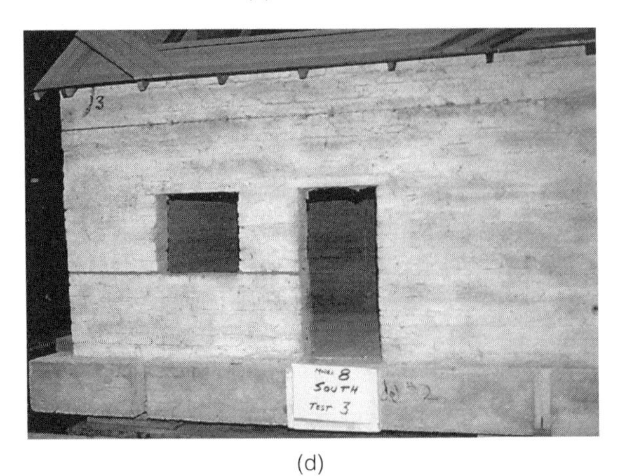

(d)

Figure 5.16
Model 8, test level III, showing (a) east wall (out of plane); (b) north wall (in plane); (c) west wall (out of plane); and (d) south wall (in plane). A few minor cracks were found during test levels II and III. More cracks were observed in the north (in-plane) wall than in the out-of-plane walls. In the north wall, cracks developed at the upper corners of the door and window. The only crack in the south (in-plane) wall was in the upper corner near the west wall. In the out-of-plane direction, the west and east walls suffered little damage. The west wall had a small vertical crack above and below the window. No damage was observed in the east wall.

(a)

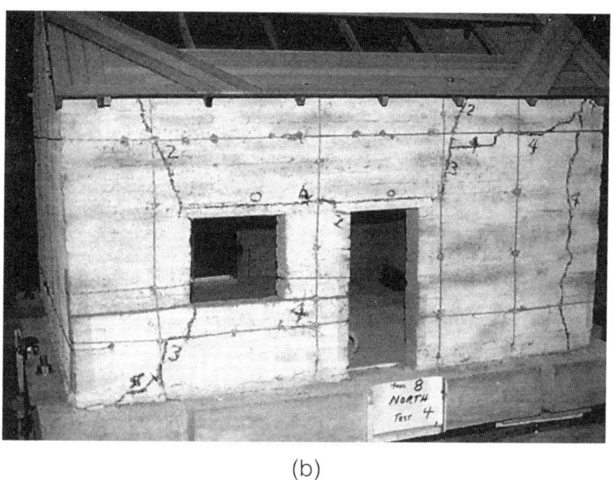

(b)

(c)

(d)

Figure 5.17
Model 8, test level IV, showing (a) east wall (out of plane); (b) north wall (in plane); (c) west wall (out of plane); and (d) south wall (in plane). In the out-of-plane direction, both the east and west walls developed vertical cracks near the center of the wall, and both had some horizontal crack damage near the base. The out-of-plane walls acted as plates with vertical yield lines at mid-span. In the in-plane direction, the north wall (straps) developed a more complete crack pattern that included a horizontal crack in the west side near the top of the wall and a vertical crack at the west end. The south wall (center cores) had a vertical crack at the west end caused by lateral movements of the west gable-end wall.

(a)

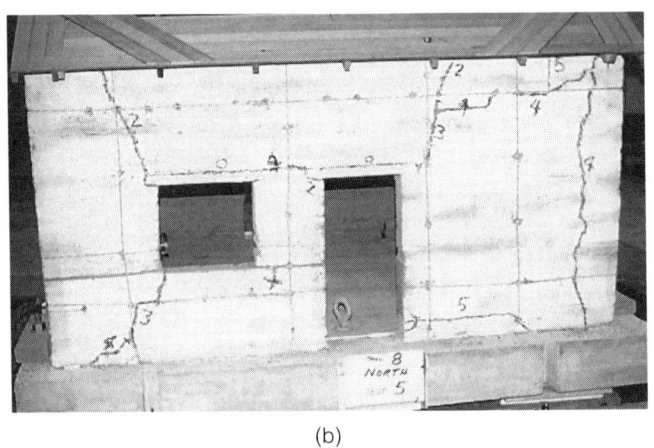

(b)

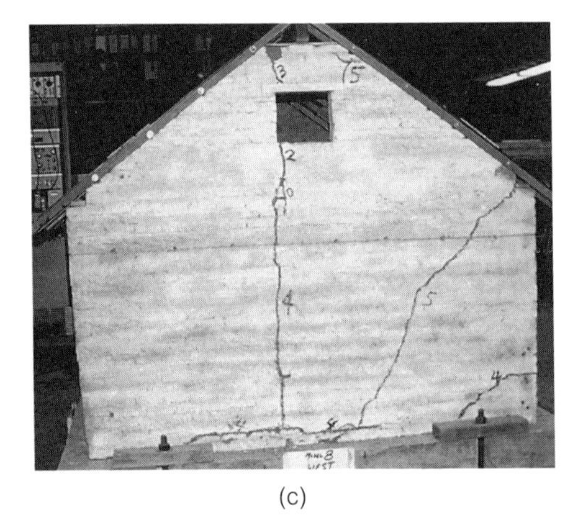

(c)

(d)

Figure 5.18
Model 8, test level V, showing (a) east wall (out of plane); (b) north wall (in plane); (c) west wall (out of plane); and (d) south wall (in plane). The in-plane walls sustained little additional damage. Minor cracks developed in the north wall, and no cracks appeared in the south wall. The out-of-plane walls developed considerably more cracks. The location of the cracks followed typical yield-line patterns. Each wall developed a diagonal crack from the upper corner on one side, extending to the center of the base of the wall. The east wall developed an additional, nearly vertical crack near the south end.

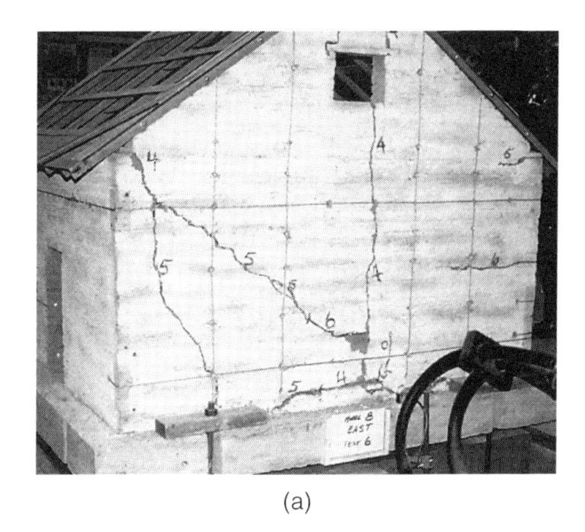

(a)

(b)

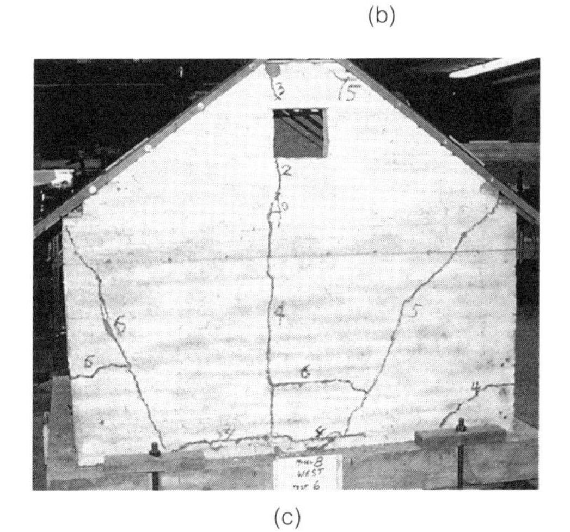

(c)

(d)

Figure 5.19
Model 8, test level VI, showing (a) east wall (out of plane); (b) north wall (in plane); (c) west wall (out of plane); and (d) south wall (in plane). Crack patterns continued to develop in the east and west walls. The west wall sustained additional diagonal cracking at the northern end and a horizontal crack near the center, located at about the level of the end of the center-core rods. The two rods near the center of the wall did not extend to the slab. In the in-plane walls, there were no additional cracks in the south wall, but further cracks appeared in the north wall. The basic crack pattern in each wall was nearly complete except for the lack of damage to the south wall.

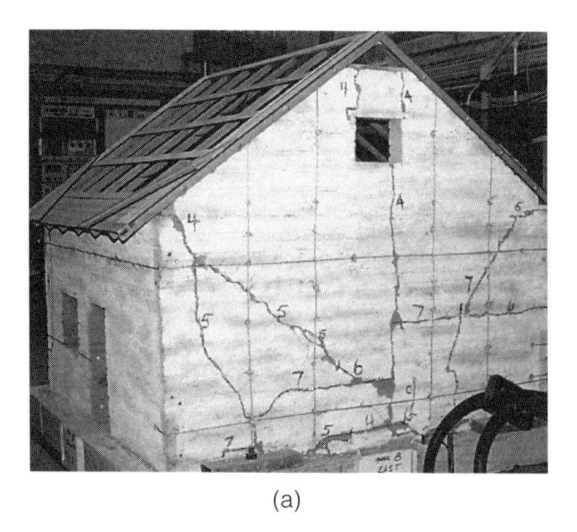

(a)

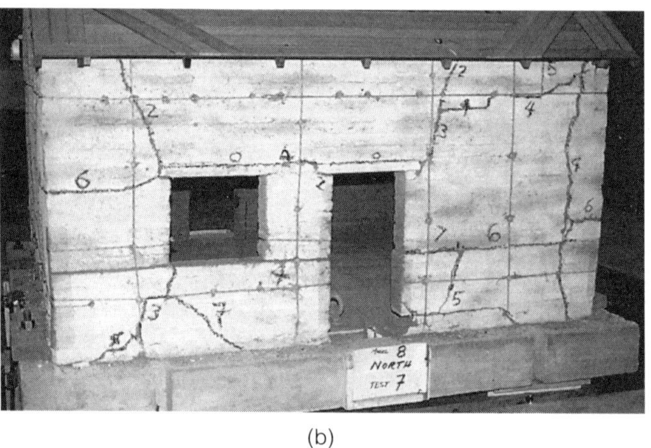

(b)

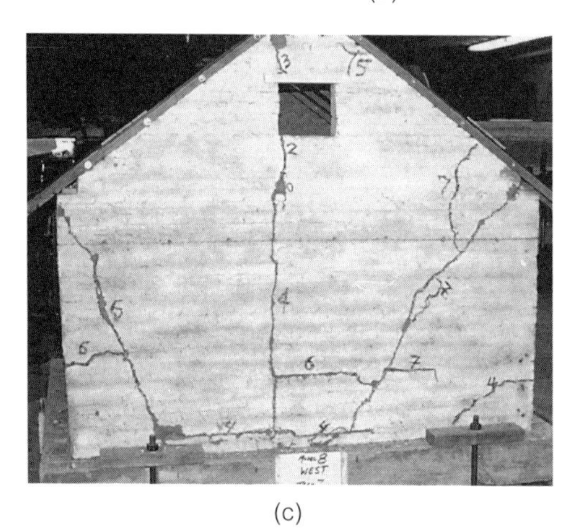

(c)

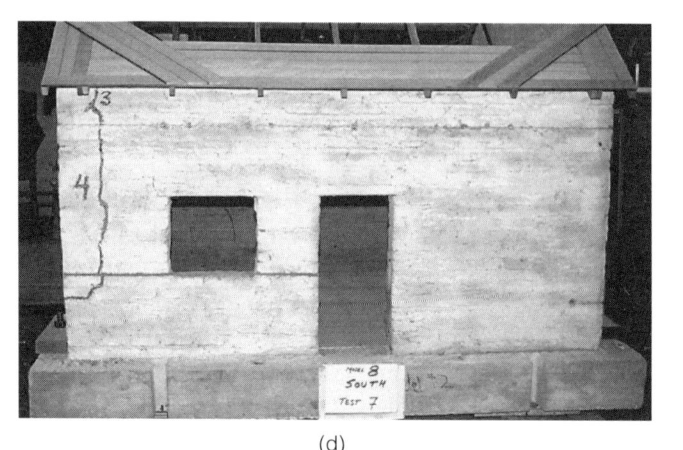

(d)

Figure 5.20

Model 8, test level VII, showing (a) east wall (out of plane); (b) north wall (in plane); (c) west wall (out of plane); and (d) south wall (in plane). At this point, the building walls had cracked into major sections that allowed them to move dynamically without sustaining substantial additional damage. Minor fragmentation of the wall continued in the north and west walls (straps), and little additional damage developed in the west wall (center core). The center cores in the south wall continued to provide adequate reinforcement that prevented cracks from occurring.

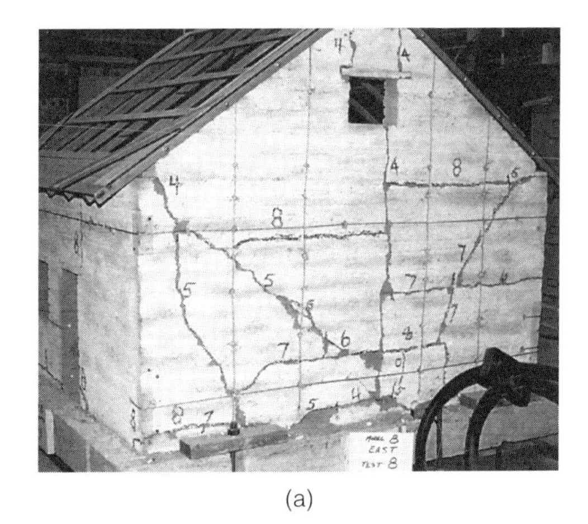

(a)

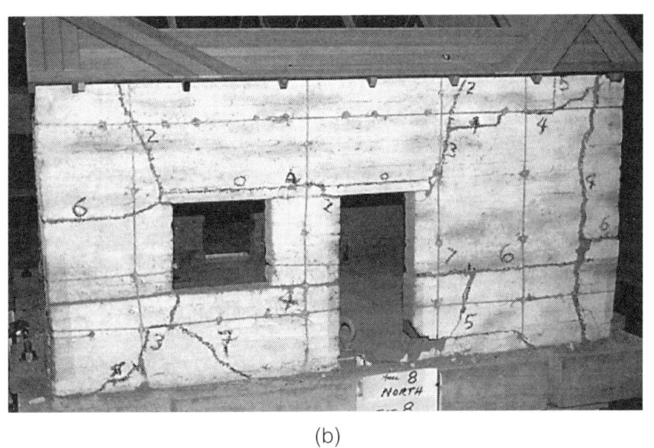

(b)

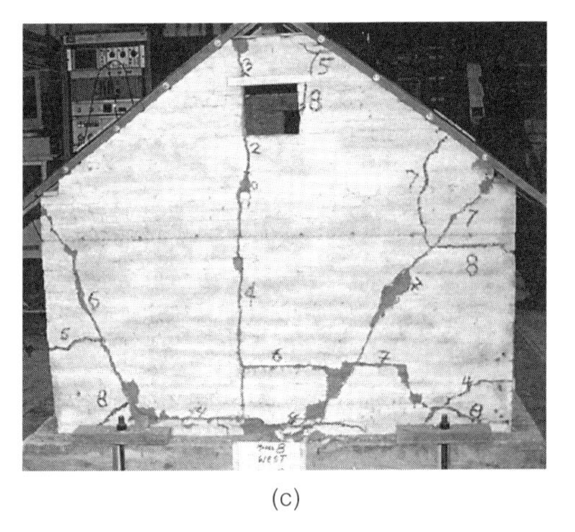

(c)

(d)

Figure 5.21
Model 8, test level VIII, showing (a) east wall (out of plane); (b) north wall (in plane); (c) west wall (out of plane); and (d) south wall (in plane). Several cracks finally developed in the south wall of this model, including a vertical crack at the door and a horizontal crack at the base of the wall below the window. The cracks that had developed in the other three walls remained fairly stable, although offsets of about 0.6 cm (0.25 in.) were observed in the north and east walls.

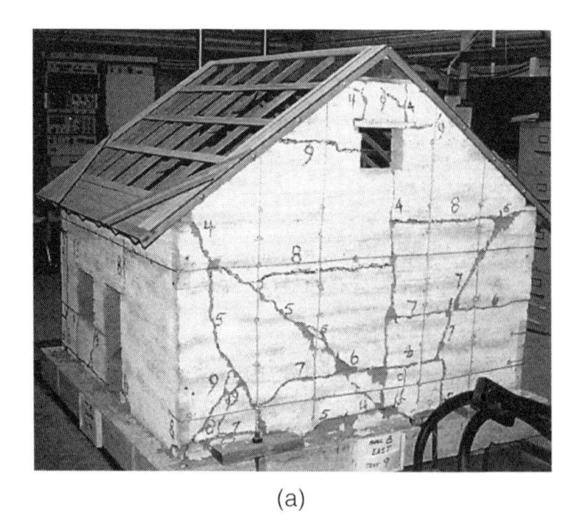

(a)

(b)

(c)

(d)

Figure 5.22
Model 8, test level IX(1), the first of two tests at this level, showing (a) east wall (out of plane); (b) north wall (in plane); (c) west wall (out of plane); and (d) south wall (in plane). Crack patterns changed little, but the substantial amount of additional spalling at crack edges and corners was indicative of the large motions to which the building was subjected during this test. The center cores in the west wall allowed the formation of major blocks with two center-core rods in each block. Damage to the south wall (with center cores) increased but was considerably less than that observed in the north wall (with straps).

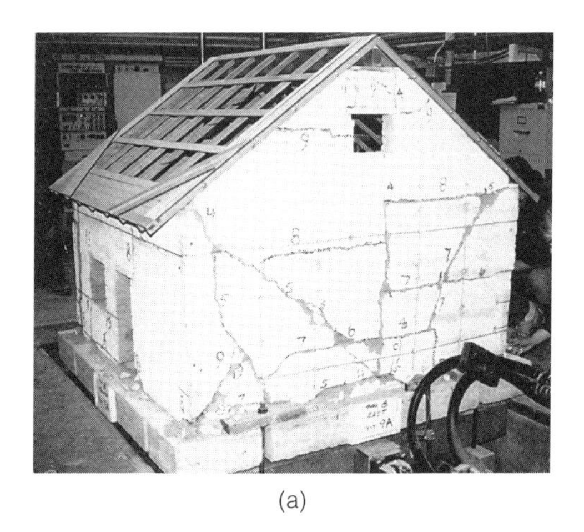

(a)

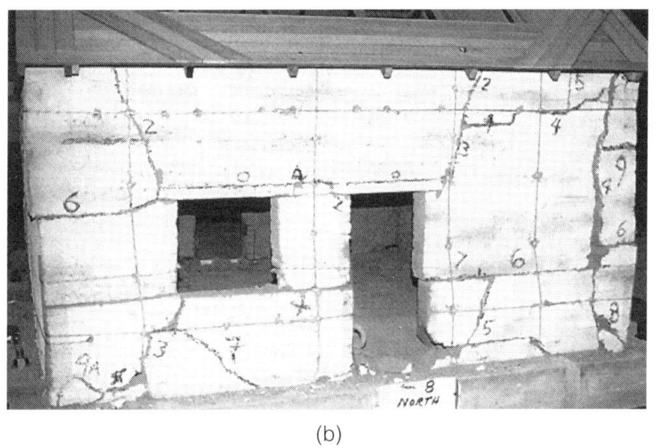

(b)

(c)

(d)

Figure 5.23
Model 8, test level IX(2), the second of two tests at this level, showing (a) east wall (out of plane); (b) north wall (in plane); (c) west wall (out of plane); and (d) south wall (in plane). Offsets at existing cracks increased to about 1.25 cm (0.5 in.). Damage to the basic panels in the south wall was still negligible. The walls were beginning to slide at the foundation level.

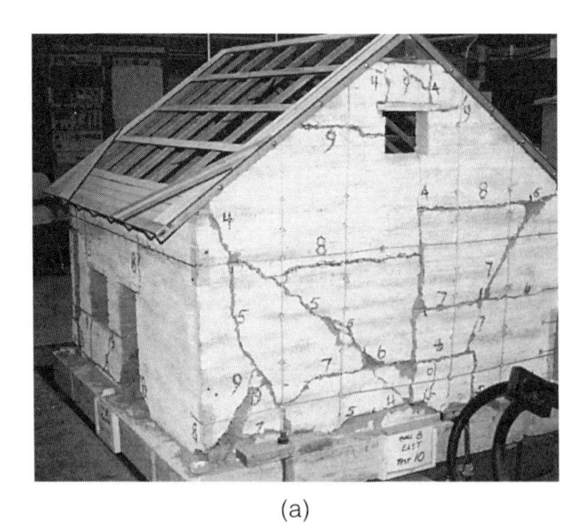

(a)

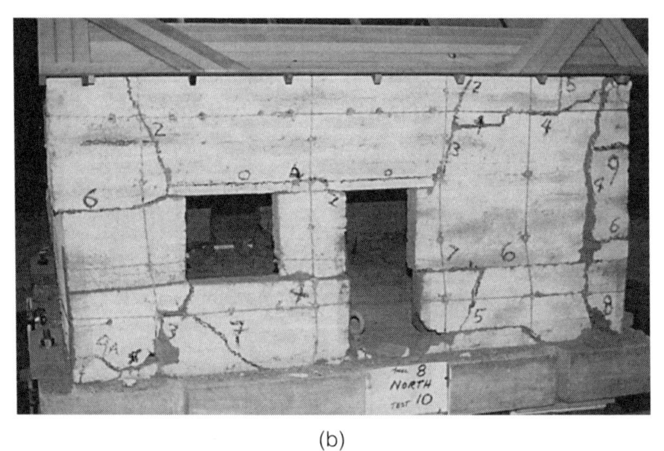

(b)

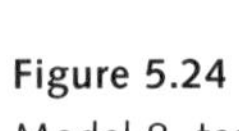

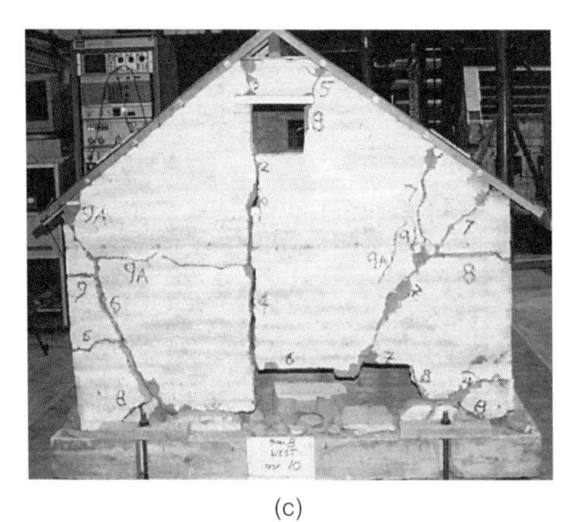

(c)

(d)

Figure 5.24
Model 8, test level X, showing (a) east wall (out of plane); (b) north wall (in plane); (c) west wall (out of plane); and (d) south wall (in plane). Crack pattern remained fundamentally the same, and no new cracks were created. The center-core rods prevented major damage from occurring in the walls in which they were installed. For the most part, cracks formed in the area just outside of the center-core rods. There were no rods or straps in the section at the west end of the south wall, which permitted an offset of more than 1.25 cm (0.5 in.) at that location. The offsets in the east and north walls increased to about 1.8 cm (0.7 in.). The straps did an effective job of maintaining stability but were less effective than center-core rods at controlling damage.

Figure 5.25
Close-up of center-core rod epoxy after removal of adobe.

the fragility of an unretrofitted adobe building. Cracks began during test levels II and III and continued to develop during test level IV. The dynamic motions of the gable-end wall became apparent during test level IV, and the model nearly collapsed during test level V. The roof rafters on both sides of the gable-end walls provided some resistance to the motion of these walls and probably were responsible for the stabilization of the model during test level V (fig. 5.27).

Both gable-end walls collapsed during test level VI (fig. 5.28). The collapse of the east wall was more complete because the upper section of the wall collapsed inward and, simultaneously, the lower section of the wall collapsed outward. The west wall collapse was incomplete. The difference in the behavior between the two walls can be attributed to the crack patterns that developed during earlier tests. There were two horizontal cracks in the east wall, and both the upper and lower sections

Figure 5.26
Section of wall panel containing center-core rod before static-loading test.

Figure 5.27
Model 9, test level V, showing (a) east wall (out of plane); (b) north wall (in plane); (c) west wall (out of plane); and (d) south wall (in plane). The dynamic motion of the gable-end walls increased substantially, and the model became nearly unstable. The slight resistance provided by the roof framing may have prevented out-of-plane collapse of the gable-end walls. A permanent offset of approximately 1.9 cm (0.75 in.) perpendicular to the plane of the wall occurred at the center of the upper horizontal crack in the east wall. A few additional cracks developed near the base of the east wall, and the vertical cracks in the west wall continued to extend downward.

of this wall collapsed. In the west wall, there was only a single horizontal crack located at a height lower than the upper horizontal crack in the west wall.

Damage to the in-plane walls was typical of that observed during previous testing. Offsets in these walls were small during all of the tests because the table motions were not strong enough to lead to significant in-plane damage. In previous research performed at Stanford, significant in-plane damage did not occur until test level VII, thus the performance of model 9 was consistent with past observations.

Summary of Test Results for Models 8 and 9

Tests on models 8 and 9 showed the dramatic effects that can result from implementing a complete, stability-based retrofit system. The performance of these models was consistent with results from the previous GSAP tests and contributed to the refining of the stability-based reinforcing techniques.

The primary goal of retrofit measures is to prevent collapse, and that goal was effectively accomplished by the system installed on model 8. A secondary goal was to minimize the extent of damage during moderate to large ground motions. The strapping method was slightly more effective at limiting permanent displacements than similar methods

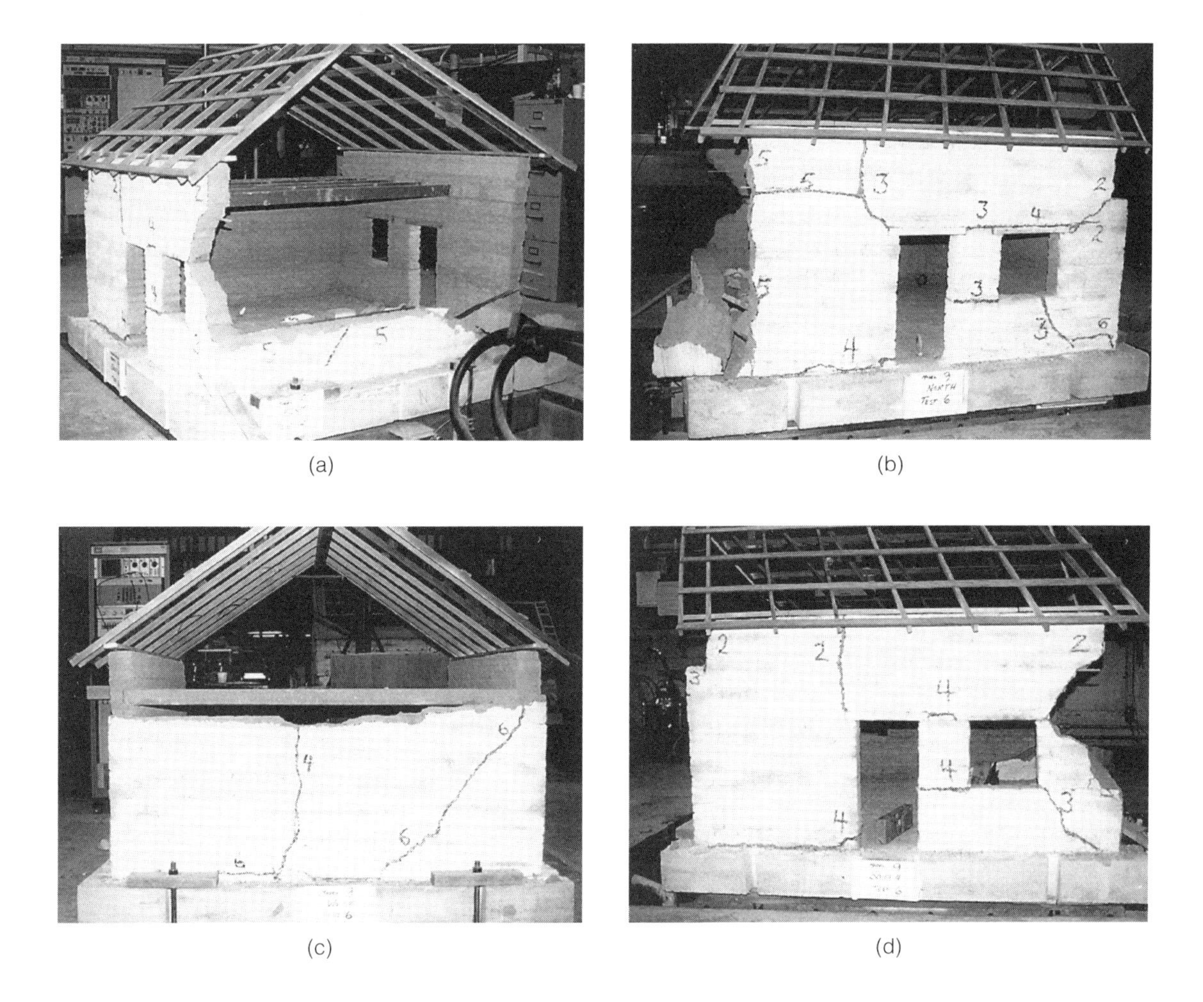

Figure 5.28
Model 9, test level VI, showing (a) east wall (out of plane); (b) north wall (in plane); (c) west wall (out of plane); and (d) south wall (in plane). Large sections of both gable-end walls collapsed. The upper section of the east wall collapsed inward at the beginning of the test, followed by the collapse of the lower section of that wall. In the latter part of the test, the upper section of the west gable-end wall also collapsed. The conditions of the in-plane walls did not worsen significantly during this test. The missing section at the east end of the south wall was attached to the section of the east wall that collapsed out of plane.

used on model 7. This difference can be attributed to the use of more closely spaced crossties and the addition of wire mesh at some locations to distribute the loading at corners under the nylon straps, which helped prevent the straps from digging into the adobe.

As mentioned, the performance of the walls that had been retrofitted with fiberglass center-core rods embedded in an epoxy grout was unexpectedly good. These walls—south (in plane) and west (out of plane)—suffered very little damage during the tests. During placement of the center-core rods, the epoxy grout was absorbed unevenly into the adobe and permitted the rods to bond effectively to the adobe material.

In summary, the important aspects of the performance of models 8 and 9 are as follows:

1. Model 9, the control model, collapsed during test level VI. In the prototype domain, the peak acceleration was approximately 0.4 g and the peak displacement was 19 cm (7.5 in.). The model nearly collapsed during the previous test.
2. The retrofit systems installed on model 8 were effective in preventing collapse of this building.
3. Vertical straps can limit the damage during very strong ground motions but have little effect on limiting crack development during moderate ground motions.

4. Center-core reinforcement can reduce damage during both moderate and large ground motions. The rods used in model 8 acted as reinforcing bars and were effective in strengthening the adobe walls.
5. The diagonal braces at the corners of the partial roof diaphragm in model 8 were effective in increasing the stiffness of the roof system over that of the roof of model 7. This roof system was designed for use in these test model buildings, however, and a similar system is unlikely to be used in a full-scale building. An existing roof will be completely sheathed with boards or plywood, which is likely to provide sufficient rigidity.
6. The saw cuts in the south wall were successful in that cracks occurred at these locations rather than radially at stress-concentration locations at the window corners. Horizontal cracks are easier to manage than cracks that are diagonally oriented.

Chapter 6
Analysis of Test Results for Models 1–9

The analysis and synthesis of the most important aspects of GSAP research on small-scale models are presented in this chapter. In the research program, tests were conducted on six simple models and three tapanco-style models that included roof and floor framing.

The first section of this chapter contains a discussion of (1) the important dynamic characteristics of the model buildings, (2) the level of damage that can be expected during different levels of ground acceleration, and (3) the effectiveness of the various retrofit measures on the building systems.

A brief summary of results of observations of damage to historic adobe buildings made after the January 17, 1994, Northridge earthquake is presented in the second section.

The final section is a synthesis of the results from both the model tests and field observations. In general, there is good correlation between damage observed in the laboratory and that observed in the field. This agreement is important for those assessing damage from past earthquakes, determining the risk of damage in the future, and designing seismic retrofits for existing historic adobe buildings.

Analysis of Model Test Results

Initial crack development (test levels II–IV)

Crack initiation began during early tests at EPGA values of 0.18 g to 0.28 g. Typically, cracks began during test level III and increased somewhat during test level IV, but full crack development did not occur until test levels V and VI. There were some clear indications of differences in crack development that were functions of the slenderness ratios of the walls.

Effect of slenderness ratios on out-of-plane performance

The test level required to cause damage to out-of-plane walls was one of the greatest differences in the performance of thick and thin walls (fig. 6.1). Thin walls (S_L = 11) are shown in figure 6.1a, b, and thick walls (S_L = 5) are shown in figure 6.1c, d. As would be expected from an elastic analysis, the crack pattern in the out-of-plane walls (fig. 6.1a) occurred first, followed by cracking in the in-plane walls. Once cracks had developed in the out-of-plane walls, the loads were quickly transferred to the

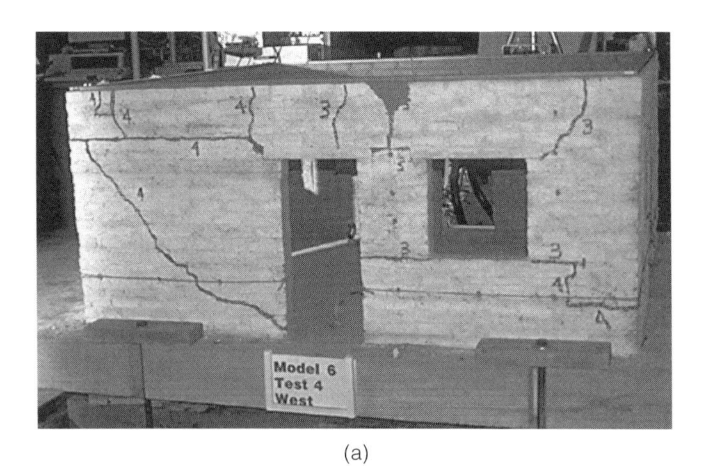

(a)

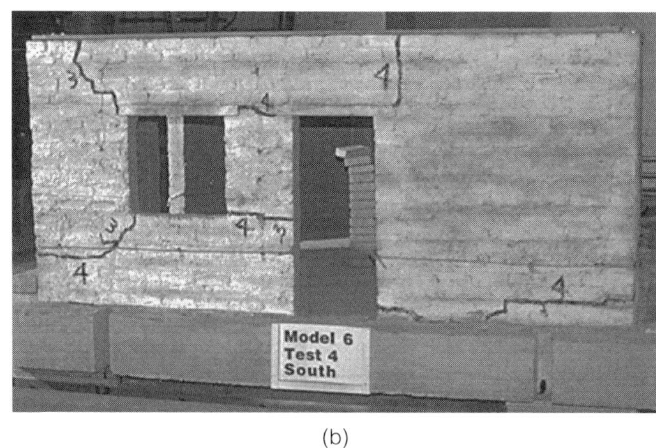

(b)

(c)

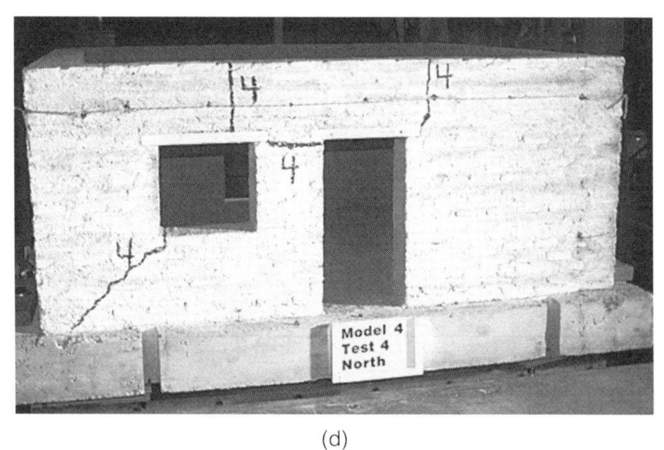

(d)

Figure 6.1
Comparison of crack initiation after test level IV for thin-walled model 6 (S_L = 11), showing (a) west wall, and (b) south wall; and thick-walled model 4 (S_L = 5), showing (c) east wall and (d) north wall.

in-plane walls through the attached bond beam, causing cracks in the in-plane walls (fig. 6.1b).

In contrast, the out-of-plane thick walls (S_L = 5) suffered very little damage during test level IV, while the in-plane walls suffered cracking at the corners of some of the openings (fig. 6.1c, d). The thick walls had much greater flexural stiffness and strength than the thinner walls and were mostly able to resist the loading with only a minor crack at the bottom of the window opening. The out-of-plane thick walls were strong enough to resist the loading and were sufficiently stiff to transfer the load to the in-plane walls, which suffered damage similar to that of the thin walls. As would be expected from an elastic analysis, the thin walls (S_L = 11) were much more susceptible to out-of-plane damage than the thick walls (S_L = 5). There was little difference in the initial crack patterns of the in-plane walls.

Effect of slenderness ratios on in-plane performance

In general, the level of damage to the in-plane walls was independent of wall thickness. A comparison of in-plane wall damage after test level VII for walls with differing slenderness ratios (S_L = 5, 7.5, and 11) is shown in figure 6.2. Damage was similar for each of these walls. The principal damage occurred at the corners of the doors and windows and at the top and bottom of the pier between the door and the window, which was typical of both laboratory and field observations.

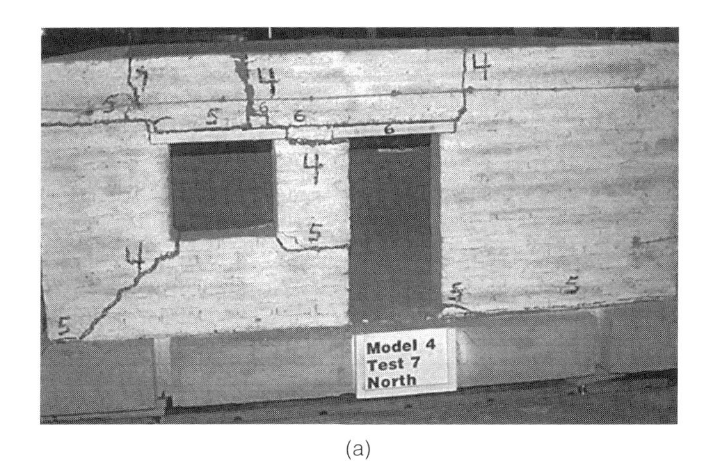

(a)

(b)

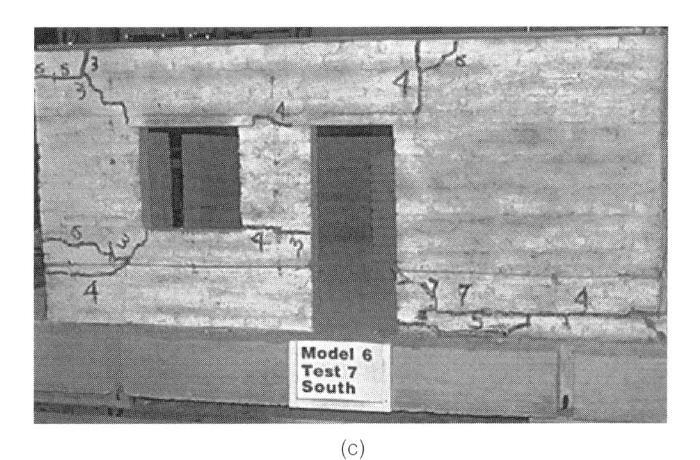

(c)

Figure 6.2
Comparison of damage of in-plane walls with differing slenderness ratios during test level VII: (a) model 4 (S_L = 5), north wall; (b) model 1 (S_L = 7.5), north wall; and (c) model 6 (S_L = 11), south wall. Similar damage was typically observed in all in-plane walls at this test level.

The performance of the walls with epoxied center-core elements was the exception to this observation. The epoxied rods acted as effective reinforcements and delayed the appearance of crack damage. Damage to the walls of model 8 (fig. 5.19) and to the walls with epoxied center cores was very slight compared to the full crack development of the walls fitted with vertical straps.

Similarities and variations in crack patterns

The crack patterns of corresponding walls were often consistent from one model to the next and in opposite walls of the same building. Nevertheless, there was one difference between similar walls that were subjected to the same input motions. The four walls shown in figure 6.3 were all thin walls (S_L = 11). The retrofit measures had a negligible effect on their elastic performance, and the resulting crack patterns had many similarities. The principal motion of the walls was out-of-plane rocking of the center panel, as shown in the whitened area in figure 6.4. In figure 6.3a–c, the crack pattern is nearly the same. This pattern is defined by a crack in the upper left corner of the wall that extends diagonally to the base of the door. From the right side, the crack starts at the upper right corner of the wall and extends through the window opening. However, in figure 6.3d, no major diagonal crack in the large left-hand panel is observed, suggesting that variability in material and construction typically exists in adobe structures, which could account for the difference.

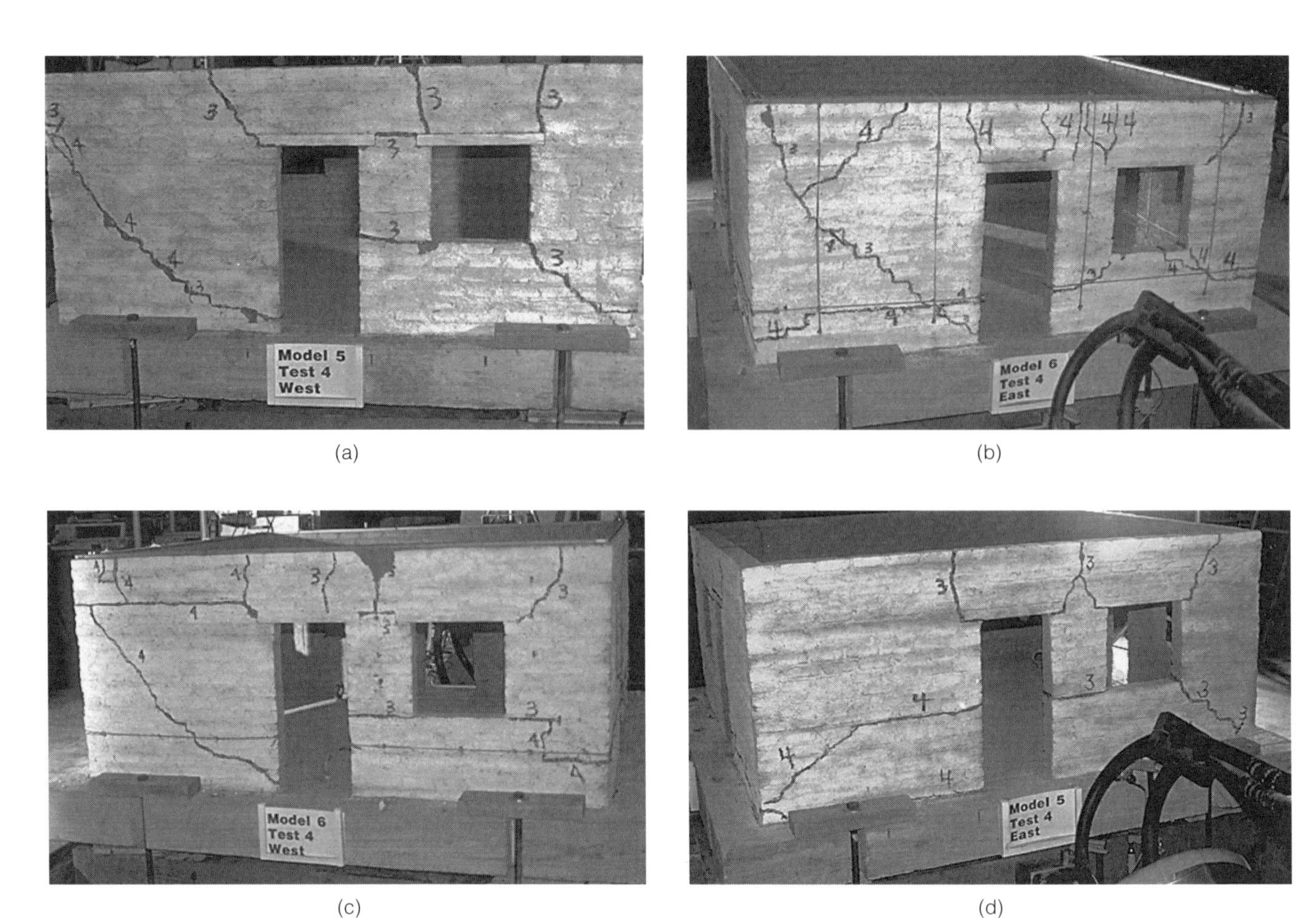

Figure 6.3
Comparison of crack patterns in out-of-plane walls with the same slenderness ratio (S_L = 11) after test level IV: (a) model 5, west wall; (b) model 6, east wall; (c) model 6, west wall; and (d) model 5, east wall.

Effect of slenderness ratio on the dynamic performance of walls with bond beams

The dynamic out-of-plane motion of the thin walls (S_L = 11) that had a wood bond beam was significantly different from that observed in the moderate and thick walls (S_L = 7.5 and 5). The thin walls (S_L = 11)

Figure 6.4
Cracked wall section of model 6, showing pattern typical of a section of wall subjected to out-of-plane rocking about its base.

easily rocked about their bases, and the principal lateral support was provided by the bond beam whose stiffness and tributary mass defined the dynamic characteristics of the walls. This behavior was not observed in the walls of moderate thickness (S_L = 7.5) with the same bond beam; the thickness of the wall did not permit easy rocking about the base, which significantly affected the dynamic motion of the walls. The out-of-plane motion at the tops of the walls was not amplified as it was in the thinner walls. This effect is difficult to observe using static images, but figure 6.5a, b shows damage indicative of this behavior. In these photographs, spalling of the adobe at the tops of the walls can be seen at the points of anchorage between the bond beam and the wall. This spalling is an indication of the large load transfer required for the bond beam to stabilize the walls. The thicker walls (S_L = 7.5), shown in figure 6.5c, d, did not have the same type of amplified motion, and spalling did not occur at the anchor bolts at the tops of the walls. The greater wall thickness also reduced the likelihood of spalling. This behavior is clearly evident in the videotaped recording of these tests.

Performance during moderate to strong seismic levels (test levels V–VII)

Complete crack development

Complete crack development typically occurred during test levels V, VI, and VII with EPGA values of 0.32 g, 0.40 g, and 0.44 g, respectively.

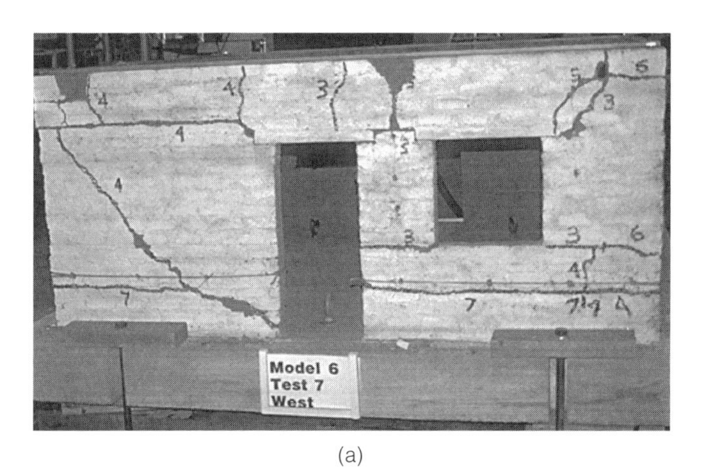

(a)

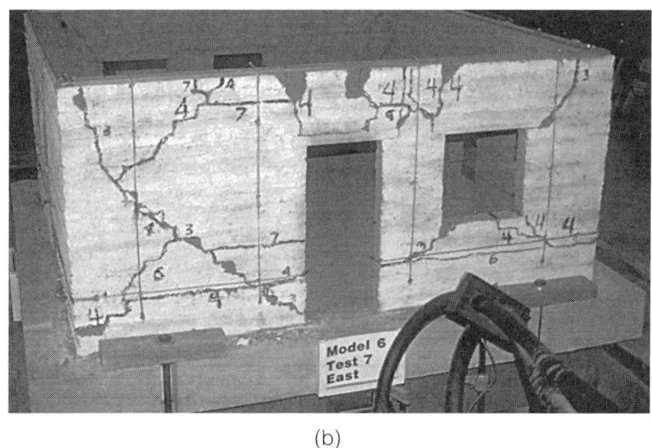

(b)

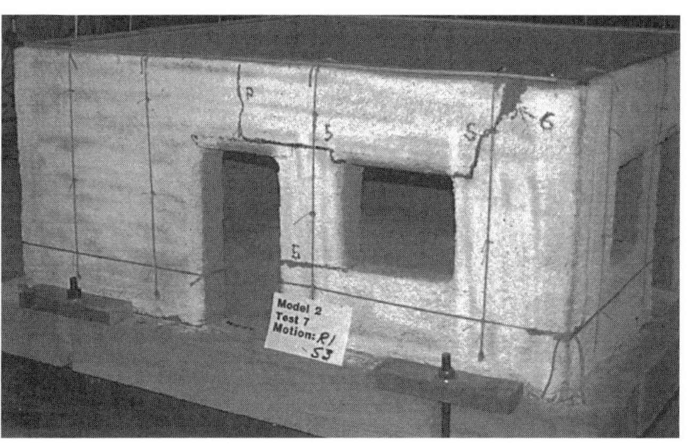

(c)

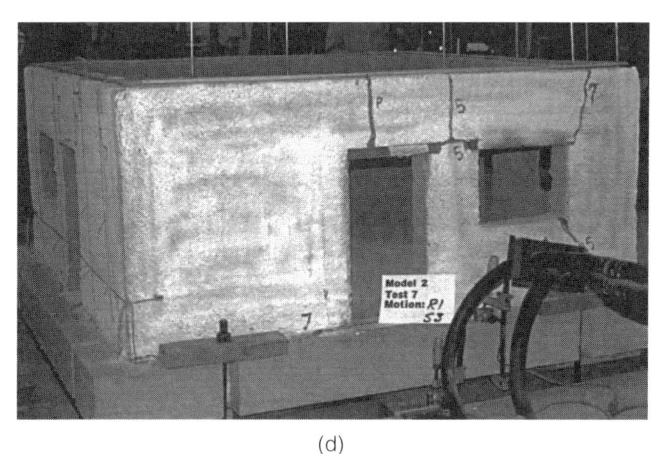

(d)

Figure 6.5
Comparison of damage after test level VII, indicative of the rocking motion of the out-of-plane walls for the thinner-walled building with bond beams, such as model 6 (S_L =11): (a) west wall and (b) east wall. Damage along the tops of thinner walls is indicative of the amplifying effect of the bond beam. Similar behavior was not observed in the thicker walls of model 2 (S_L = 7.5): (c), west wall and (d) east wall.

Minor cracking occurred after test level VII, but these cracks were primarily the result of further fracturing of the major cracked wall sections.

Unretrofitted models

The unretrofitted model 5 building collapsed almost completely during test level V (fig. 3.10). Unretrofitted model 9 nearly collapsed during test level V. In-plane walls suffered severe damage, offsets, and partial collapse during test level VI (fig. 5.28).

Effectiveness of retrofit measures on performance during test levels VI and VII

The effectiveness of the retrofit measures during test level VI is shown by the performance of the tapanco-style buildings, as illustrated in figure 6.6. Unretrofitted gable-end walls, tested out of plane, collapsed during test level VI (fig. 6.6a). The retrofitted gable-end walls suffered crack damage, but the model was very stable (fig. 6.6b). Damage to the in-plane unretrofitted walls was similar (fig. 6.6c), and those in the retrofitted structure (fig. 6.6d) suffered some additional cracks not observed in the unretrofitted model. Presumably, these were due to the additional load transferred through the partial diaphragms and upper horizontal straps.

Figure 6.6

Comparison of tapanco-style control model 9 and retrofitted model 8 after test level VI: (a) east wall of model 9 (unretrofitted) vs. (b) east wall of model 8 (retrofitted); and (c) north wall of model 9 (unretrofitted) vs. (d) north wall of model 8 (retrofitted).

(a) (b)

(c) (d)

Performance during very strong seismic levels (test levels VIII–X)

Test levels VIII, IX, and X, with EPGA levels of 0.48 g, 0.54 g, and 0.58 g, respectively, were challenging for the model buildings. The effectiveness of a retrofit was demonstrated by successful performance during these three tests. The lighter retrofit measures failed during one of these tests, while the more robust and resilient retrofit solutions performed well after several repetitions of test level X.

Comparison of retrofit measures during test levels VII and VIII

The light retrofit system used on the thin south and west walls of model 6, which included a bond beam and vertical local ties, was adequate to withstand the motions of test level VII, but the out-of-plane wall collapsed during test level VIII (fig. 6.7a). By comparison, the opposite walls of the same model were stable during this test (fig. 6.7b), a result that can be attributed to the effectiveness of the vertical straps. Similarly, moderately thick walls with a bond beam and center-core rods also behaved well during test level VIII (fig. 6.7c).

Comparison of retrofit measures during test level IX

Test level IX was even more of a challenge to some of the lightly retrofitted model buildings. A large section of the in-plane thin wall with bond beam and local ties collapsed during this test (fig. 6.8a). Again, the

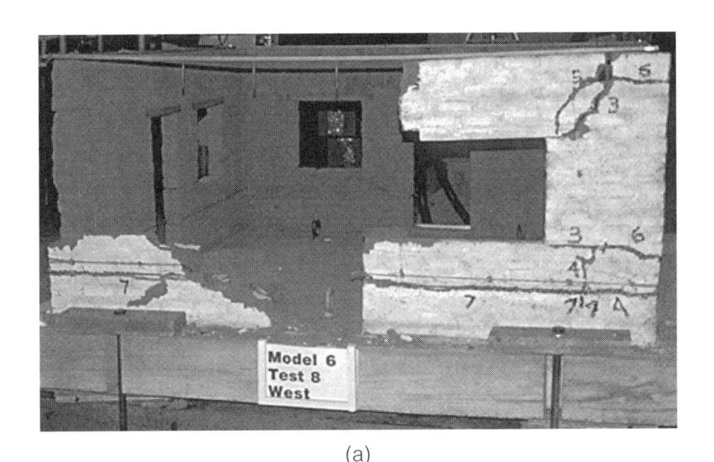

(a)

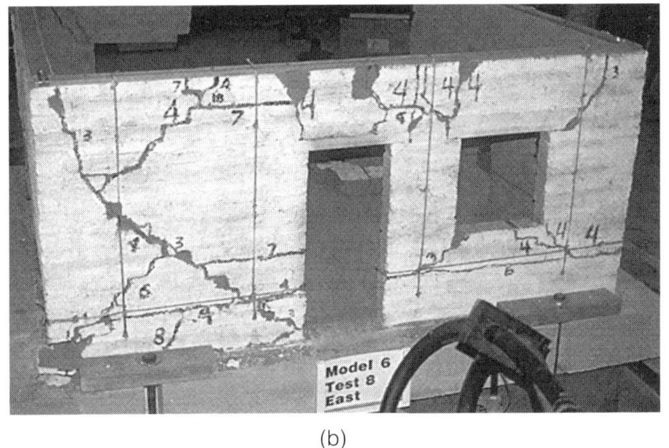

(b)

(c)

Figure 6.7
Comparison of out-of-plane performance after test level VIII of (a) lightly retrofitted walls (model 6, west wall) vs. (b) more complete retrofits that included vertical straps (model 6, east wall) or (c) center cores (model 2, east wall).

(a)

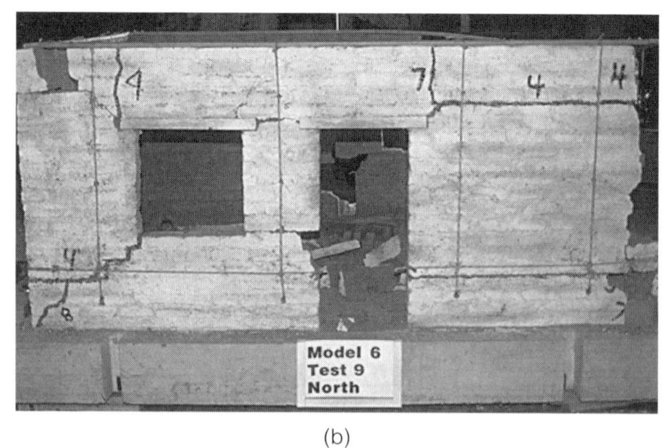

(b)

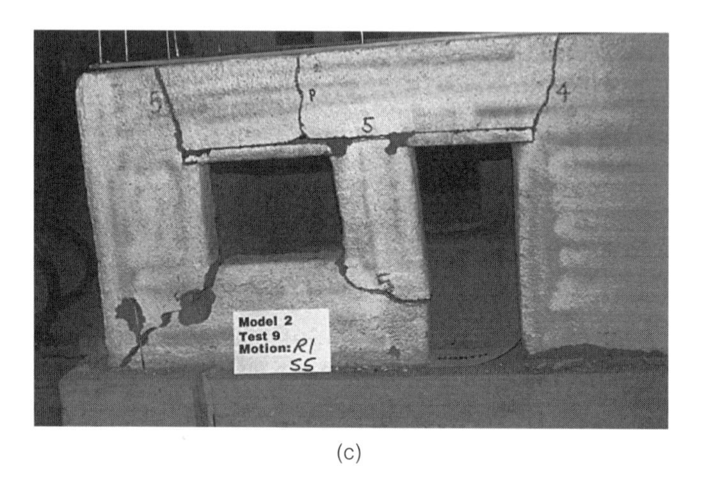

(c)

Figure 6.8
Comparison of in-plane performance after test level IX of (a) lightly retrofitted models (model 6, south wall) vs. (b) more complete retrofits that included vertical straps (model 6, north wall) or (c) center cores (model 2, north wall).

light retrofit system used on this thin wall (S_L = 11) was not adequate to provide sufficient resilience to withstand the demands of this major dynamic event.

The addition of vertical straps instead of local ties improved the performance of the thin walls. The condition of the thin wall with a bond beam and vertical straps after test level IX is shown in figure 6.8b; the cracked wall is still stable and generally in good condition. The condition of the two in-plane walls shows the difference between the use of local ties (fig 6.8a) and the use of vertical straps (fig. 6.8b). Good behavior was also observed in the moderate walls with a bond beam and center cores during test level IX (fig. 6.8c).

Increased effectiveness of bond beams when combined with vertical straps or center cores

The behavior of the model buildings with bond beams and little additional retrofitting demonstrated the inadequate resistance of a bond beam when used without other retrofit measures. In the previous Stanford testing (Tolles and Krawinkler 1989), one model (S_L = 7.5) was retrofitted with only a wood bond beam. While the beam was more than sufficient to transmit the loading from the out-of-plane walls to the in-plane walls, the bond beam alone was not sufficient to restrain the movements of the cracked block sections of the walls. Relative displacements along cracks accumulated as the tests progressed, and collapse occurred. The model with the bond beam collapsed at the end of test level VIII. In the GSAP research discussed here, the light retrofit system using a bond beam and local ties (S_L = 11) behaved similarly to the model with only a bond beam; the out-of-plane wall collapsed during test level VIII, and the in-plane wall collapsed during test level IX.

Model behavior as a function of wall slenderness and retrofit measures
The behavior of the model buildings during the most extreme tests was a function of both the thickness of the walls and the type of retrofit measures installed. With very thick walls (S_L = 5), only minimal retrofit measures were needed to prevent instability. All four thick walls with horizontal straps survived, with only the addition of local ties to prevent instability of the pier between the door and window openings. The out-of-plane east wall, however, was nearly unstable after test level X (fig. 6.9a). Walls of moderate thickness (S_L = 7.5), when retrofitted with the same horizontal straps, survived test level IX but collapsed during test level X (fig. 6.9b). This collapse may have been prevented by additional through-wall ties along the length of the upper horizontal straps. The performance of these walls indicates that additional vertical elements may be necessary for the survival of moderate to thin walls during the severest ground motions.

Figure 6.9
Comparison of out-of-plane wall performance in lightly retrofitted models after partial collapse: (a) model 4 (S_L = 5), east wall, test level X; (b) model 1 (S_L = 7.5), east wall, test level X; (c) model 6 (S_L = 11), west wall, test level VIII; and (d) model 6 (S_L = 11), east wall, test level X.

(a)

(b)

(c)

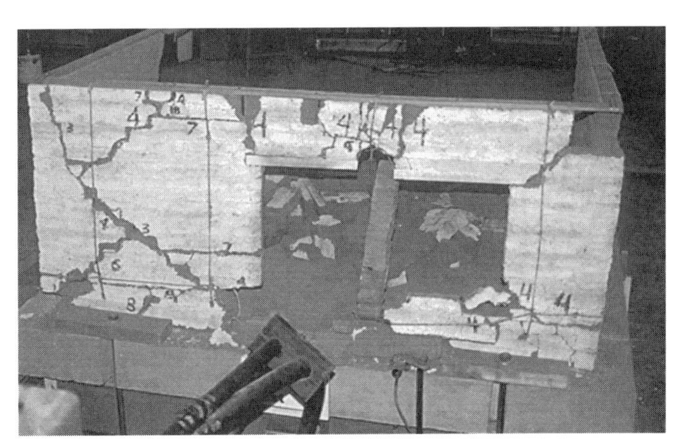
(d)

The performance of a thin wall with a bond beam and local ties showed that modest improvements to the retrofit systems on a thin-walled building still may not be sufficient. The out-of-plane walls collapsed during test levels VIII and IX, respectively (fig. 6.9c, d). These two walls had a bond beam, lower horizontal straps, and local ties in critical locations.

The other two walls of the same model behaved considerably better with the addition of vertical straps. Nevertheless, during test level X, the out-of-plane east wall lost the pier between the door and window even with the vertical strap (fig. 6.9d). This type of damage was not observed in other buildings with thicker walls when vertical straps or center-core rods were installed.

Effectiveness of a complete retrofit system demonstrated by the performance of tapanco-style models 7–9

The performance of the tapanco-style models 7–9 demonstrated convincingly the effectiveness of the installed retrofit systems. Figures 6.10 and 6.11 show the behavior difference of out-of-plane and in-plane walls, respectively, of the retrofitted models after the final test. As noted earlier, the unretrofitted model 9 (fig. 6.6a) collapsed near the beginning of test level VI. The performance of the retrofitted models 7 and 8 that survived test level X was, therefore, extremely successful by comparison.

Figure 6.10
Comparison of the out-of-plane wall performance of tapanco-style models with differing retrofit systems after test level X: (a) model 7, east wall; (b) model 7, west wall; (c) model 8, east wall; and (d) model 8, west wall.

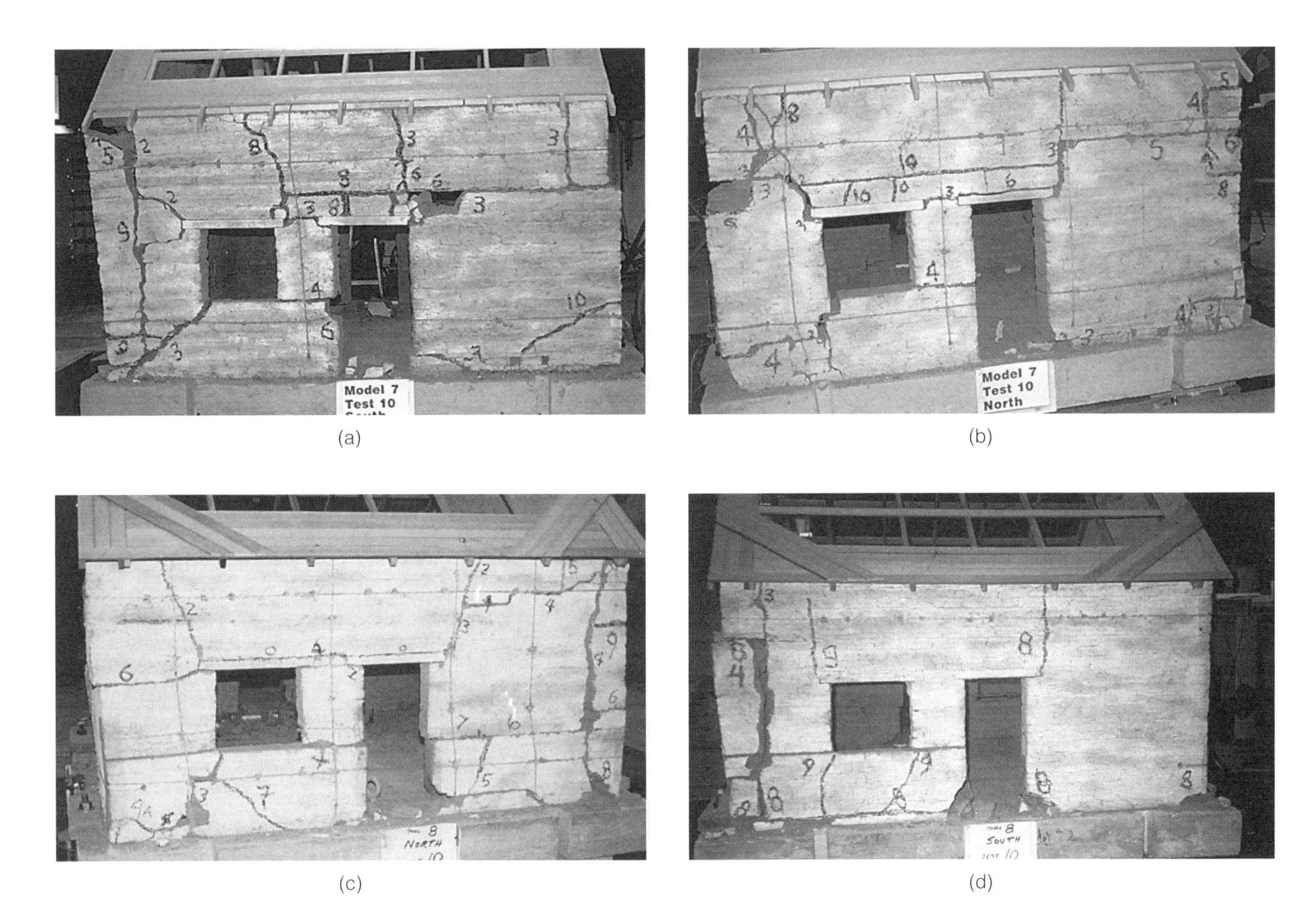

Figure 6.11
Comparison of the in-plane wall performance of tapanco-style models with differing retrofit systems after test level X: (a) model 7, south wall; (b) model 7, north wall; (c) model 8, north wall; (d) model 8, south wall.

Comparison of out-of-plane, gable-end walls

The gable-end walls of the tapanco-style models were all tested out of plane. In addition to the partial diaphragms on each building, the minimum retrofits were upper and lower horizontal straps. The only gable-end wall that did not have lower horizontal straps was the west wall of model 8 (fig. 6.10d), which had center-core rods. In model 7, the east wall (fig. 6.10a) had only upper and lower straps, yet the thick walls (S_L = 5) remained stable. There was an approximately 3.8 cm (1.5 in.) offset at the mid-height horizontal crack, because there was little restraint across this crack plane.

Vertical straps provided additional restraint that minimized the extent of dynamic and permanent crack offsets. In model 7, the performance of the west wall (fig. 6.10b) was somewhat better than that of the east wall as a result of the vertical straps that provided restraint for block movements along horizontal and diagonal cracks.

Effect of more closely spaced crossties in tapanco-style models

The retrofit system used on model 8 (S_L = 7.5) was modified slightly to try to decrease the amount of offset at horizontal cracks. The east wall (fig. 6.10c) had vertical straps, and the wall crossties were spaced closely together to limit the relative displacement between adjacent blocks. The displacements in the gable-end wall were limited to 1.25–1.9 cm (0.5–0.75 in.) in the model domain, which was an improvement over the performance of model 7 but was not as dramatic as that achieved by the installation of center-core rods.

Effect of center-core rods in tapanco-style models

The performance of the west wall of model 8 showed the extent to which center-core rods can reduce the offset across cracks during out-of-plane ground motions (fig. 6.10d). This wall had center-core rods that extended from the top to almost the base of the wall. Most of this wall was retrofitted with epoxied center-core rods and sustained minimal crack damage. There were no measurable offsets across center-core rods. Extensive damage to this wall occurred near the base in the area where the center-core rods did not extend. Most of the energy was dissipated in the areas outside of the center-core rods. If the center-core rods had extended into the foundation, there may have been additional damage in the upper part of the wall since there would have been no clear area for energy dissipation.

In the in-plane direction, center-core rods were also most effective at preventing permanent offsets. Figures 6.10 and 6.11 show the damage to the retrofitted tapanco-style models at the end of the testing sequence. In general, the cracks opened to approximately 1.9 cm (0.75 in.) in each of the walls except where epoxied center-core rods were installed. The horizontal straps restrained the development of permanent offsets in the walls but still allowed displacements of the blocks. Stiffer straps may have been used to provide greater restraint against damage, but two factors would need to be taken into account: (1) a failure plane might develop just above the strap that would permit slippage, or (2) the strap may dig into the adobe material and loosen, even with a stiffer strap. To try to control the latter, a wire-mesh screen was added at the exposed corners. The straps were still able to dig into the adobe at the through-wall holes. Minimal damage was observed in the in-plane wall of model 9, where the vertical center-core rods were used (fig. 6.11d).

Summary of performance of model buildings

The retrofit measures used on in-plane and out-of-plane walls and the results of the final tests on models 1–9 are summarized in tables 6.1 and 6.2.

Initial cracking and moderate damage

Typically, cracks were initiated during test level III or IV at EPGA values of 0.23 g and 0.28 g, respectively. Cracks continued to develop during test levels V and VI; after these tests, the principal cracked block sections were observed to have developed. The full development of the cracked block sections then determined the nature of seismic performance during the succeeding tests (test levels VII and higher).

Severe damage and collapse

Unretrofitted models collapsed during test level V or VI. These levels had EPGA values of 0.32 to 0.40 g, respectively, and, more important, the peak displacements were approximately 15.2–17.8 cm (6–7 in.) in the

Table 6.1
Models 1–6: Summary of results for wall pairs

Model no. and walls[a]	Collapse level	Principal retrofit measures	Comments
5	VII	None (control model)	Complete collapse
4 NE	No collapse	Upper horizontal strap	Basically stable with substantial block offsets
1 NE	X	Upper horizontal strap	Out-of-plane collapse that may have been prevented by more closely spaced crossties
4 SW	No collapse	Upper and lower horizontal straps	Basically stable with substantial block offsets
1 SW	No collapse	Upper and lower horizontal straps	Close to collapse during final test
6 SW	VIII	Bond beam, lower horizontal straps, and local ties at piers between the door and windows	Collapse of out-of-plane west wall during test level VIII; collapse of most of south wall during test level IX
6 NE	No collapse	Bond beam, lower horizontal straps, and vertical straps	Out-of-plane walls near collapse; center pier dislodged
2 NE	No collapse	Bond beam and center-core rods	Stable behavior in all tests
2 SW	No collapse	Bond beam, lower internal horizontal straps, and vertical straps	Stable behavior in all tests
3 NE	No collapse	Bond beam, lower internal horizontal straps, and vertical center-core rods	Stable behavior in all tests
3 SW	No collapse	Bond beam, lower internal horizontal straps, and vertical center-core rods	Stable behavior in all tests

[a]NE = north and east walls; SW = south and west walls. East and west walls were tested out of plane; north and south walls were tested in plane.

Table 6.2
Models 7–9: Summary of results for wall pairs

Model no. and walls[a]	Collapse level	Principal retrofit measures	Comments
9	VI	None (control model)	Complete collapse of gable-end walls
8 NE	No collapse	Partial wood diaphragms—upper strap at attic-floor level, lower straps, and vertical straps	Stable behavior in all tests
8 SW	No collapse	Partial wood diaphragms—upper strap at attic-floor level, lower straps, and vertical center-core rods; no lower strap on west wall	Stable behavior in all tests
7 NE	No collapse	Partial wood diaphragms—upper strap at attic-floor level, lower straps, and vertical straps	Stable behavior in all tests
7 SW	X	Partial wood diaphragms—upper strap at attic-floor level and lower straps; no vertical straps	Partial collapse of south (in-plane) wall during test level X

[a]NE = north and east walls; SW = south and west walls. East and west walls were tested out of plane; north and south walls were tested in plane.

prototype domain. Thin walls (S_L = 11) were approximately 28 cm (11 in.) thick, moderately thick walls (S_L = 7.5) were 41 cm (16 in.) thick, and thick walls (S_L = 5) were 61 cm (24 in.) thick.

Very high accelerations will not destabilize adobe walls if the frequency is high and the displacements small. If the displacements are small, the base input may have increased accelerations, but without large displacements it is nearly impossible to destabilize the thick walls that are typical of most historic adobe buildings.

Lightly retrofitted models were able to withstand somewhat higher test levels. Thin walls with a bond beam and local ties collapsed out of plane during test level VIII, and the in-plane wall with the same retrofit system collapsed during test level IX. The thick walls with horizontal straps nearly collapsed during test level X, while the moderately thick walls with the same retrofit system did collapse during test level X.

Other models with more complete retrofit systems showed greater resilience to extended, strong table-motions. The addition of either vertical straps or center-core rods greatly improved overall structural performance at the higher test levels. Walls with horizontal and vertical retrofit measures continued to behave well through repetitions of test level X, and no indications of imminent collapse were observed. The performance of the thin east wall of model 6 survived test level X, although the pier between the door and window was dislodged.

Summary of Field Observations after the Northridge Earthquake

Based on the information generated during the survey of damage to historic adobe buildings after the 1994 Northridge earthquake (Tolles et al. 1996), estimates were made regarding expected damage to historic adobe buildings as a function of peak ground acceleration (PGA). The conclusions presented in this section are based on observations made following this and other earthquakes.

Figure 6.12 shows the relationship between the PGA level and damage for well-maintained historic adobes. Ground acceleration in the range of 0.1 g to 0.2 g PGA seems to be required to initiate damage in such buildings. At this level of ground motion, cracks begin to form at door and window openings and at the intersections of perpendicular walls. At the Southern California adobe sites studied where PGA was in this range, the Miguel Blanco Adobe, San Marino, was undamaged; the Purcell House, San Gabriel, was slightly damaged; the Plaza Church, Los Angeles, was also slightly damaged; and the Centinela Adobe, Los Angeles, experienced slight to moderate damage. These adobes were well maintained and had not been reinforced or retrofitted.

At PGA levels of approximately 0.4 g, the damage becomes more extensive. In the cases of the De la Osa Adobe, the Andres Pico Adobe, and the Del Valle Adobe (see chapter 1), the damage to walls was extensive throughout these structures.

Preexisting conditions affect the observed damage level greatly. In figure 6.12, all data on buildings with preexisting water and

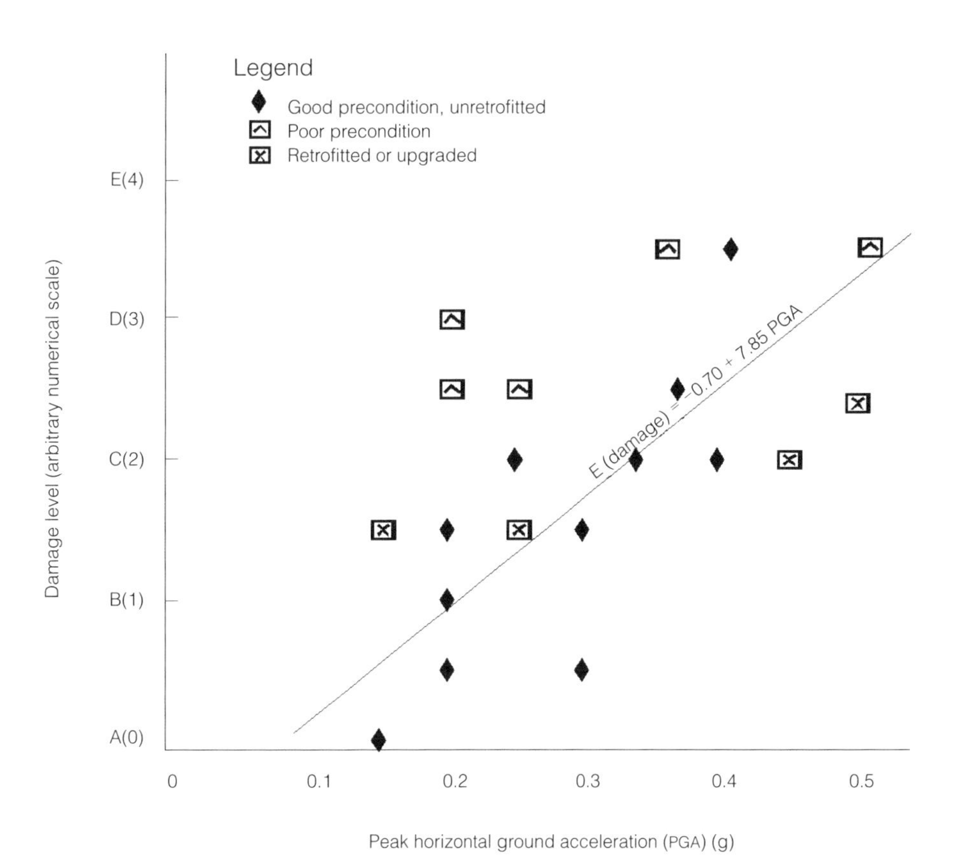

Figure 6.12
Observed damage level vs. peak horizontal ground acceleration for historic adobe buildings after the 1994 Northridge earthquake.

earthquake damage are found above the expected damage line for well-maintained adobes, indicating that severe damage can be expected for buildings in poor condition. Even for ground motions of moderate intensity (0.1–0.2 g), poorly maintained buildings are likely to suffer substantial damage.

Comparison of Laboratory Results and Field Observations

Comparison of laboratory tests and field observations show good correlation. The important observations can be summarized as follows:

1. Below 0.15 g, little or no damage will be observed in a well-maintained adobe building.
2. Starting around 0.2 g, minor cracks will be observed, though damage may be much more severe in a previously damaged structure.
3. From 0.25 to 0.35 g, damage will become increasingly severe. The number and length of cracks will increase throughout the structure, and some permanent offsets may be observed.
4. Above 0.35 g, severe damage is likely to occur throughout the structure, and wall instability may be observed.

These observations are valid for a typical historic adobe building that has been well maintained, has thick or moderately thick walls, and is of average strength. The presence of higher-strength adobe material may delay the onset of cracking. Low-strength adobe or previously damaged walls will have a negative impact on performance, and thinner walls will be more likely to collapse. Nevertheless, these general categories can help determine the effects of larger seismic events. A graphic representation of these basic concepts is shown in figure 6.13, which is based on observations of buildings affected by the 1994 Northridge earthquake.

Many of the damage typologies observed in the field after the Northridge earthquake were also observed in the 1:5 scale model buildings. Typical out-of-plane wall failures were observed in numerous historic adobe buildings, and the crack pattern resulting from out-of-plane flexural failure was observed quite often. Gable-end wall failure was also a typical pattern observed in the field and in the laboratory. The out-of-plane collapse of load-bearing walls was observed at only one site. This failure was not the result of overturning but was caused by moisture damage at the base of the wall.

At corners, vertical and diagonal cracks in the models were observed and were similar to those observed in the field, except in-plane shear cracks occurred less frequently. Mechanical tests of the 1:5 scale adobe brick assemblies showed that shear strength was proportionately higher compared to compressive or flexural strengths. Therefore, in-plane shear cracks are less likely to occur than vertical or diagonal cracks.

In general, the overall crack patterns observed in the model buildings were similar to those found in historic adobe buildings after the Northridge earthquake. The nature of the post-elastic behavior is largely dependent on the locations of the crack patterns and the resulting movement of the cracked wall sections.

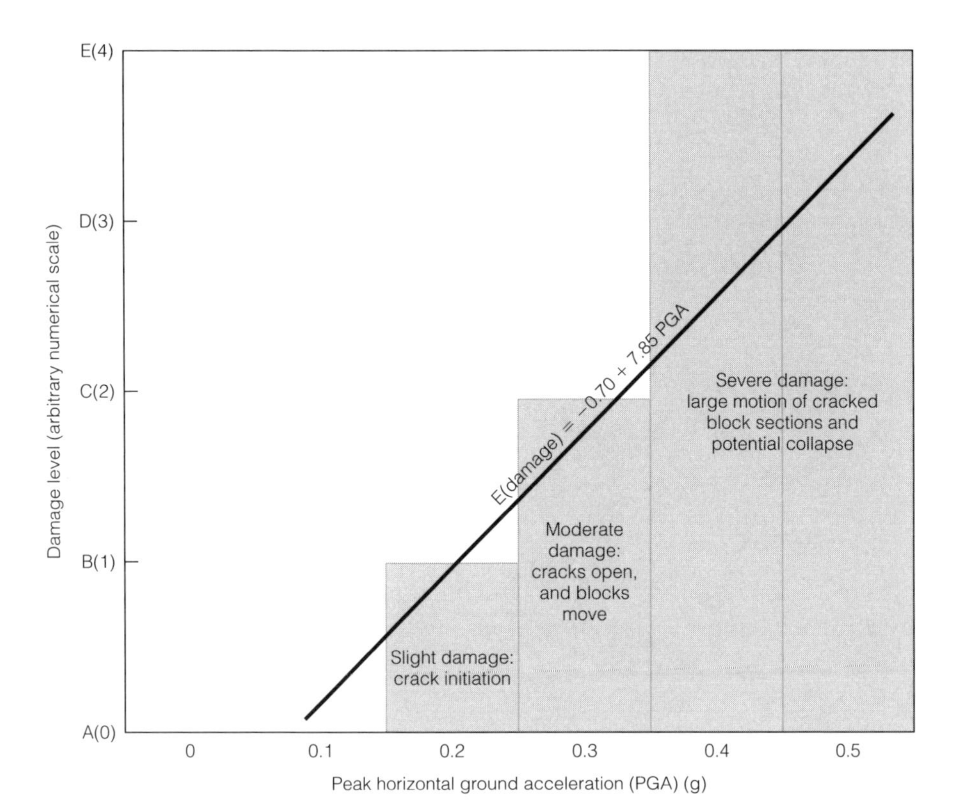

Figure 6.13
Damage level vs. peak horizontal ground acceleration for historic adobe buildings after the 1994 Northridge earthquake.

Chapter 7

Description of Tests for Models 10 and 11

The discussion of model similitude in chapter 2 pointed out that the only forces not modeled accurately were the gravitational forces that cause compressive stresses in the walls. For single-story structures, these stresses are very low, especially in the upper portions of the walls, where damage from seismic motion is usually most severe.

When the structure is damaged and becomes inelastic, the accuracy of the model is more difficult to assess. Overturning of individual walls is not properly modeled because gravity forces resist overturning in severely cracked walls. Also, sliding along cracks is not accurately simulated in models that do not include the simulation of gravitational forces. Resistance to sliding is proportional to the vertical stresses, which are smaller because gravity loads are not fully simulated. Resistance is also affected by the increased strain rates of the models. These problems are not believed to be of primary importance if the objective of the model test is to study the global response characteristics of the adobe buildings, as was the case for the 1:5 scale models. Overturning of walls in reduced-scale models may occur at slightly lower levels, and the frictional resistance along cracks may be somewhat different, but the response of the model buildings still contains the global characteristics of prototypes and can provide useful information on conceptual behavior and on comparisons of improvement techniques.

To address the possible effects of scale, however, two additional 1:2 scale model buildings, model 10 and model 11—unretrofitted and retrofitted, respectively—were constructed and tested using procedures similar to those used on the 1:5 scale models. The principal objectives of these tests were as follows:

1. To compare the performance of the 1:2 scale models with that of the 1:5 scale models in an attempt to understand the limitations of small-scale testing
2. To compare the dynamic performance of the retrofitted model with that of the unretrofitted control model
3. To acquire quantitative measurements and documentation of (a) the dynamic behavior of the models for general analysis, and (b) the loads in the retrofitting elements to allow for sizing of these elements in future design applications

In the 1:5 scale models, only 20% of the gravity load was simulated, compared to that in the prototype building. In the 1:2 scale models, the gravity loads were 50% of those in the prototype. The effect of this difference should be considered, and where there is a difference in the behavior between the 1:5 scale models and the 1:2 scale models, a slightly larger effect would be expected in the prototype.

The three basic response categories that should be characterized with respect to the effect of the scaling parameters are (1) the elastic response range, (2) the initiation and development of cracks, and (3) the post-elastic performance.

Neglecting gravity loads has a negligible effect on the prediction of the elastic behavior of one- or two-story adobe buildings. In fact, when performing analytical studies of a building of this type, the gravity loads are neglected because the compressive stresses are low and the displacements are small.

The development of cracks is also minimally affected by gravity loads. The walls of a masonry building are stiff, and therefore cracks will develop even when displacements are small. When displacements are small, gravity loads have little effect on crack development. Once the cracks in a wall of an adobe building have become fully developed, however, gravity loads may have an effect. The types of motion that are affected by gravity loads can be grouped into three categories: (1) out-of-plane rocking and overturning, (2) sliding along *horizontal* cracks where additional gravity loads increase resistance to sliding and therefore tend to *decrease* the amount of sliding, and (3) movement along *diagonal* cracks where additional gravity loads *increase* the amount of sliding.

The results of the testing program are termed *conservative* if the performance of the prototype buildings is considered to be better than that of the model buildings. The results are termed *unconservative* if the performance of the prototype is considered to be worse than that of the model buildings.

Two factors affect out-of-plane rocking and overturning: one that decreases the risk of overturning and one that increases the risk. First, the resistance to overturning of a freestanding wall is a function of the mass of the wall. If the mass of the model wall is less than that of the full-scale wall, the model wall is more likely to overturn. Second, if the mass of the wall is large enough to cause spalling at the location where cracks have developed, then the width of the wall at the crack will decrease, and the wall is more likely to overturn. The effect of spalling may make a full-scale wall more likely to overturn. These two factors have opposite effects, and the final result is not clear. If no spalling occurs, then the results of the tests on out-of-plane rocking and overturning will be conservative. But spalling may not occur in the smaller models, whereas it might occur in the prototype.

Sliding along *horizontal* or *nearly horizontal* cracks also has two opposing factors that may affect the results. On the onc hand, a greater vertical load increases resistance to sliding, and therefore a larger-scale building will have greater resistance to sliding than a smaller one. On the other hand, a reduced-scale model uses input motions that have a reduced time scale and higher frequencies. This reduced time scale increases the resistance to sliding in smaller-scale models and decreases the resistance in larger-scale models.

Movement of a wall section along a *diagonal* crack will be exacerbated by greater vertical loading because the gravity loads will increase the amount of diagonal cracking. The resistance to sliding will be less in a large-scale model due to the lower frequencies, as mentioned. In this case, both effects will increase the movement along diagonal cracks more in large-scale than in small-scale buildings. Therefore, test results will be unconservative for this type of damage.

In summary, the three types of damage affected by changes in scale and the results of these changes are as follows:

1. Overturning is dependent on two factors that have opposing effects. Crushing of adobe is usually not observed to a significant degree in the field except on walls damaged by moisture. Therefore, the smaller mass of the reduced-scale models can lead to conservative test results.
2. Sliding along horizontal cracks is probably dominated by the mass of the wall as opposed to the frequency of the motion. Again, based on this assumption, the test results will be conservative.
3. Sliding along diagonal cracks has two effects that are complementary. Therefore, the results of the tests on reduced-scale models will *not* be conservative. The results of this type of motion will be more severe in the full-scale prototype structure than in the small-scale models.

In addition to the objective of assessing the effects of increased scale on the types and locations of failure, the goal of the large-scale tests was to obtain quantitative measurements of the dynamic behavior of the models (i.e., displacement, stresses in the straps, and g-loading) that could be used for the analysis and design of retrofit elements in real buildings.

Overview of Tests

The tests on the large-scale (1:2) models were performed on the seismic simulation shaking table at the Institute of Earthquake Engineering and Engineering Seismology (IZIIS) of the University "SS. Cyril and Methodius" in Skopje, Republic of Macedonia. The models and testing parameters were basically the same as those used for the 1:5 scale tapanco models with the addition of thirty-two channels of instrumentation for measurement of peak maximum acceleration (accelerometers); displacement of the structure (linear potentiometers, clip gauges, and differential transformers); as well as stresses in the straps (load cells).

Two model buildings were tested: an unretrofitted control and a retrofitted version in which two adjacent walls had vertical and horizontal straps, and the other two adjacent walls had center-core rods embedded in epoxy grout. The buildings had wood attic floors and roof systems that were essentially the same as those of models 8 and 9.

Like models 1–9, the two large-scale models were subjected to a series of simulated earthquake motions based on the N21E component of the 1952 Taft, California, earthquake. Because the capacity of the IZIIS shaking table was exceeded during test VIII, however, higher level tests were not conducted.

Macedonian adobes had compositions and properties that were somewhat different from those of the California adobes. Thus, an extensive series of tests was carried out to determine adobe material properties and the shear and compression characteristics of isolated-wall test elements. Adobe compositions were formulated that were similar to those used for the 1:5 scale models, and the adobe walls tested had similar but lower strength properties.

Only a brief description of the model properties, test procedures, and analysis of the results is given here. A more complete description of the materials tests; the design and construction of the models; the shaking-table test facility, location of model instrumentation, and test procedures; and the detailed numerical test results are given in IZIIS Report 96-36 (Gavrilovic et al. 1996). A detailed analysis of the data obtained on models 10 and 11 is given in an internal GSAP report (Tolles and Ginell 2000).

Materials Tests

Because modeling theory required that the characteristics of the adobe materials for the models be as close as possible to those of the prototype, extensive physical testing of local Macedonian clays and wall sections was carried out. The strongest adobe material found near Skopje, however, turned out to be somewhat weaker than the adobe used in the small-scale tests. Since the difference in strength affects only crack initiation, the use of a weaker adobe was expected to have little effect on post-elastic dynamic behavior. Also, the dynamic motions of the small-scale models were stronger than those of the large-scale models during the early part of the test sequence when cracks were developing. Therefore, the lower adobe strength was compensated somewhat by the lower excitation of the model by the IZIIS shaking table.

Dynamic Testing Procedures

The strong-motion portion of each test lasted 10 seconds except for test VIII, which lasted 3.6 seconds. The capacity of the shaking table was exceeded in this test, thus the scaling factor was reduced to 1.5 (instead of 2) for test VIII* (in place of test IX). A list of test parameters is given in table 7.1.

Table 7.1
Test parameters

Test level	IZIIS test no.	Maximum base acceleration[a] (g) Model 10	Model 11	Maximum base displacement[a] mm	in.	Scaling factor of original earthquake record
I	1	0.056	0.060	25	1.00	2
II	2	0.110	0.111	51	2.00	2
III	3	0.168	0.170	76	3.00	2
IV	4	0.234	0.232	102	4.00	2
V	5	0.307	0.284	127	5.00	2
VI	6	0.392	0.384	184	6.25	2
VII	7	0.501	0.472	191	7.50	2
VIII	8[b] (model 10)	0.686		Not available	Not available	2 (short record)
	7*[b] (model 11)		0.603	Not available	Not available	2 (short record)
VIII*	8* (model 11)		0.377	254	10.0	1.5[c]

[a]All dimensions in the prototype (full-scale) domain. Displacements and accelerations are multiplied by 2.
[b]Each of these records exceeded the capacity of the earthquake simulator, which automatically stopped operating only a few seconds into the test.
[c]Required a less compressed record to avoid exceeding the capabilities of the earthquake simulator. This test had higher displacements but lower accelerations.

Model Buildings

Models 10 and 11 were built to a scale of 1:2 (S_L = 7.5) and had identical plans, dimensions, and wall layouts. The only difference was that model 11 was retrofitted. The models were erected on reinforced concrete foundations, and the adobe structure was doweled to the foundation to minimize slipping. The models were aged indoors for 80 days before testing to allow for complete drying. Dimensional and construction details of the unretrofitted model are shown in figures 7.1–7.4.

The retrofit measures consisted of

- incorporation of steel center-core rods in the west and south walls;
- application of vertical nylon straps on both sides of the east and north walls;
- application of horizontal nylon straps at two levels around the building; and
- installation of partial wood diaphragms in the attic floor and roof.

The west and the south walls were retrofitted with four vertical center-core rods placed at regular intervals (figs. 7.5 and 7.6). During building construction, 3 cm (1.2 in.) diameter cavities were formed in these walls using plastic tubes. After completion of the model, ribbed steel rods,

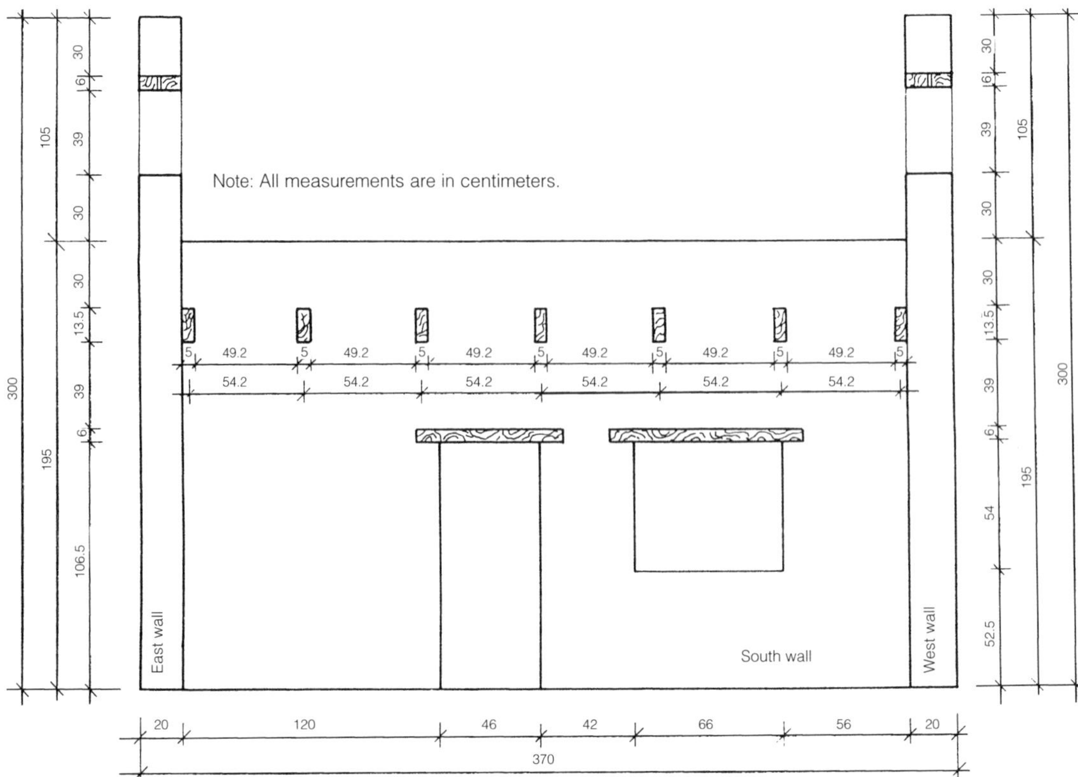

Figure 7.1
Model 10 section parallel to north-south walls.

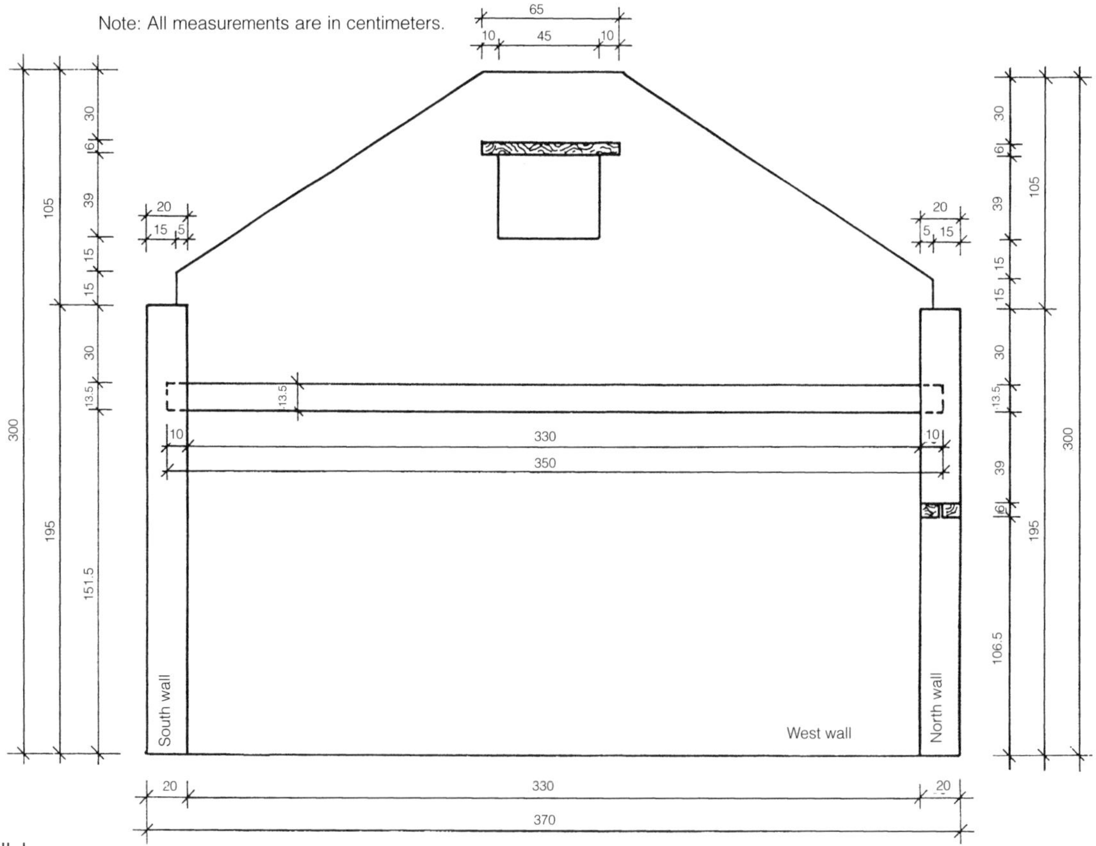

Figure 7.2
Model 10 section parallel to east-west walls.

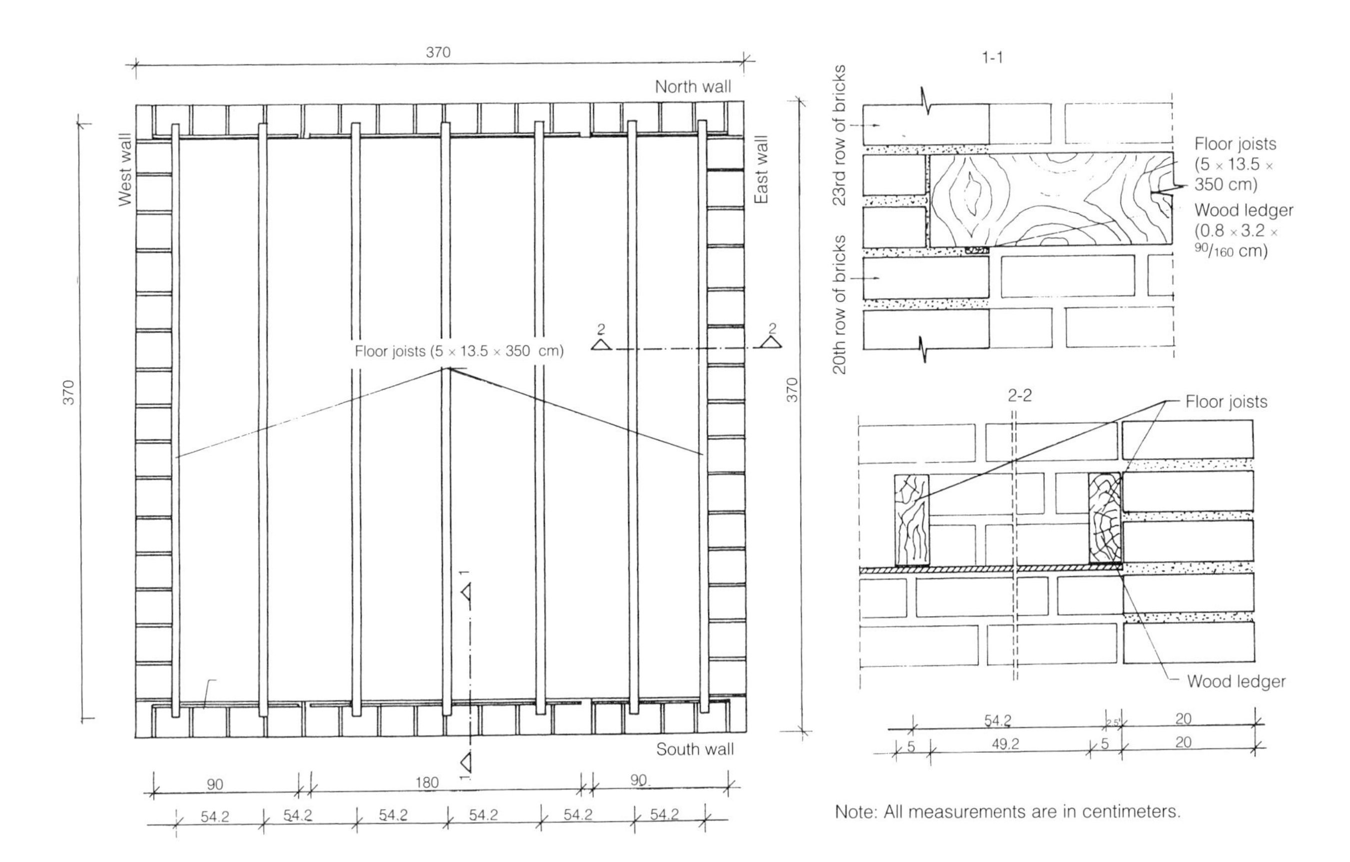

Figure 7.3
Model 10 plan at the level of the floor joists.

14 mm (0.55 in.) in diameter, were placed in these cavities and the space around them was filled with epoxy grout. The grout was a modified epoxide resin consisting of a bisphenol-A resin and a diethylene tetramine (DETA) hardener plus fillers (calcium carbonate), pigments, and additives for achieving fluidity (paraffin oil). The central core rods touched only the foundation structure. A mixture of epoxide and dry adobe was used up to a height of 20 cm (4.6 in.) to center the core rods at the bottom of the 3 cm (1.2 in.) diameter hole.

The retrofit measures installed in the east and the north walls were woven, cylindrical cross-section nylon cord straps placed at regular intervals. The straps had a measured breaking strength of 987 kg cm^{-2} (14, 010 psi). Four vertical straps were used for each wall (figs. 7.7 and 7.8). These vertical straps were located on both sides of the walls. The straps went over the tops of the walls and through drilled holes at the bases of the walls. Small-diameter, solid nylon cable ties were used as crossties to connect the straps on both sides of the wall.

Two horizontal nylon straps were installed on each of the four walls (figs. 7.5–7.8). The upper horizontal strap was located at the attic-floor line, and the lower horizontal strap was located just below the bottom of the window. The strap at the attic-floor line ran around the perimeter of the building and was attached to the floor system, as shown in figure 7.9. The lower horizontal strap was located on both sides of each wall section. The characteristics of the nylon straps used for retrofitting are given in table 7.2, which shows the tensile strength and the corresponding strain, Δ1/1 (%), as functions of applied stress for a 40 cm (16 in.) sample length.

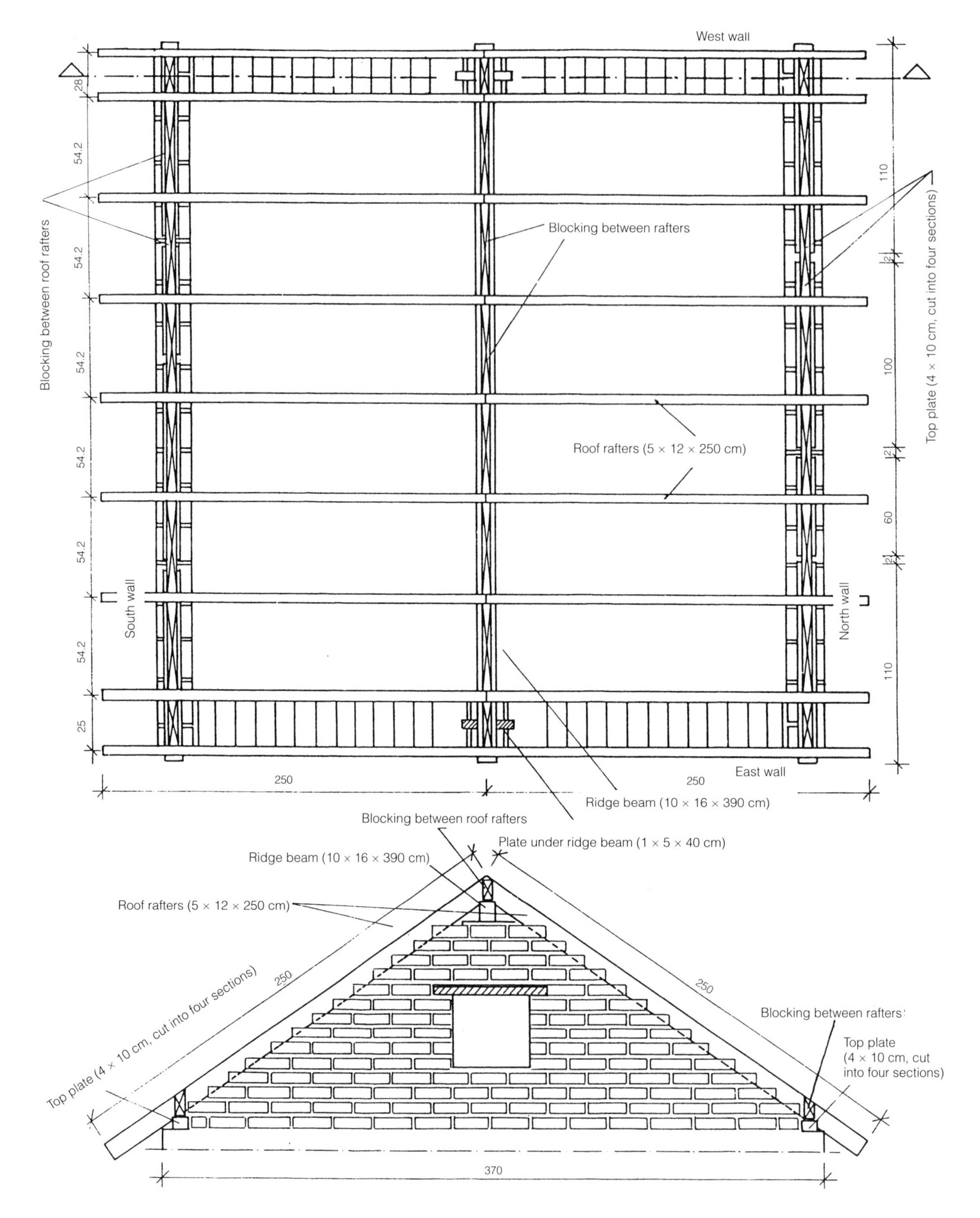

Figure 7.4
Model 10 roof framing plan.

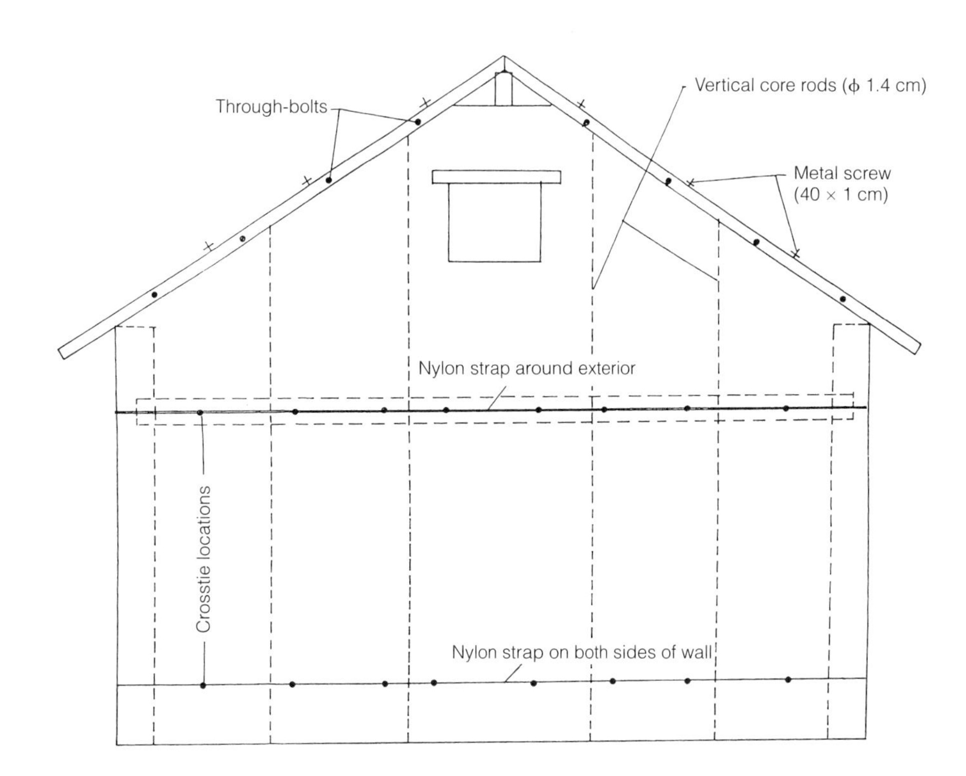

Figure 7.5
Model 11 elevation of west wall.

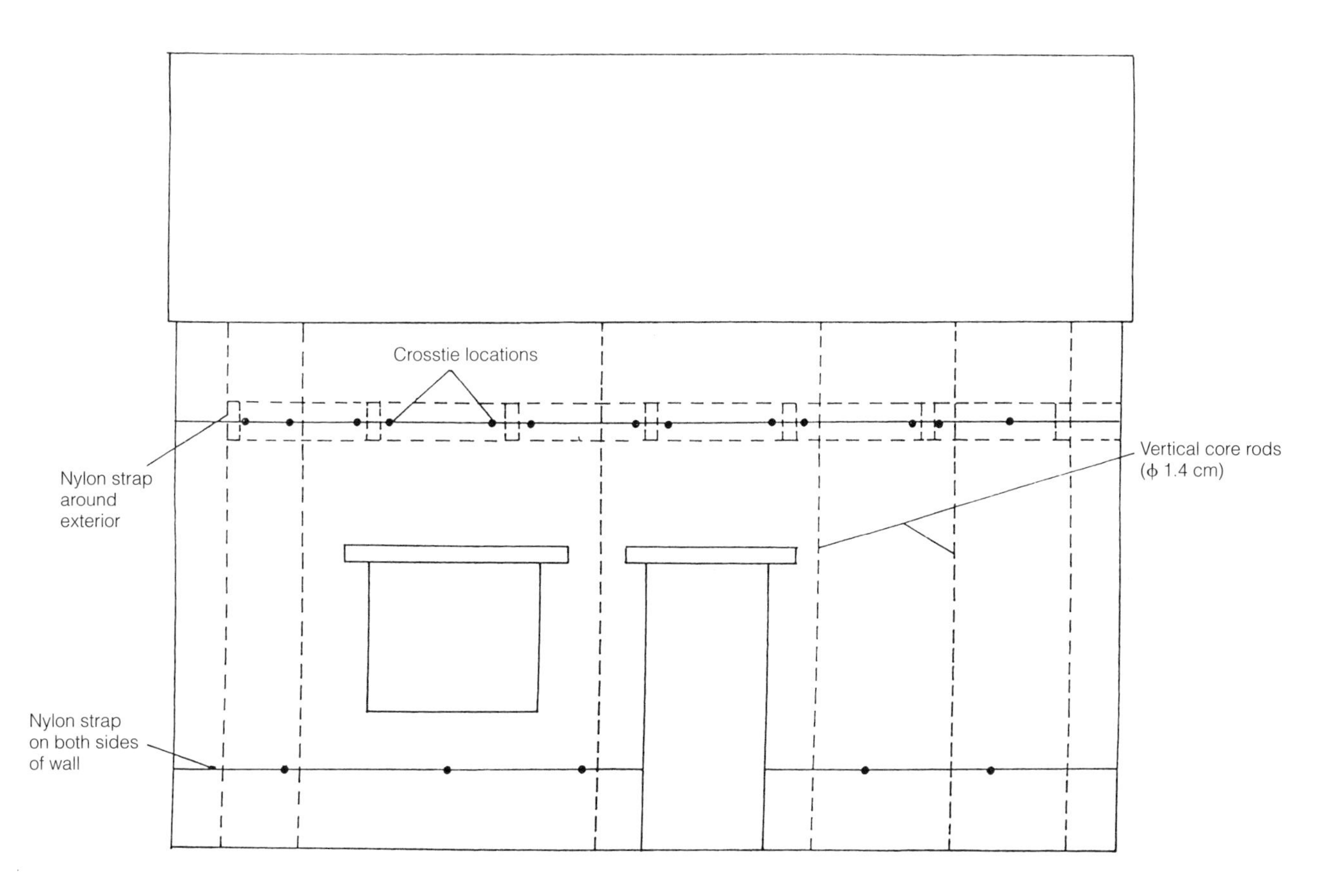

Figure 7.6
Model 11 elevation of south wall.

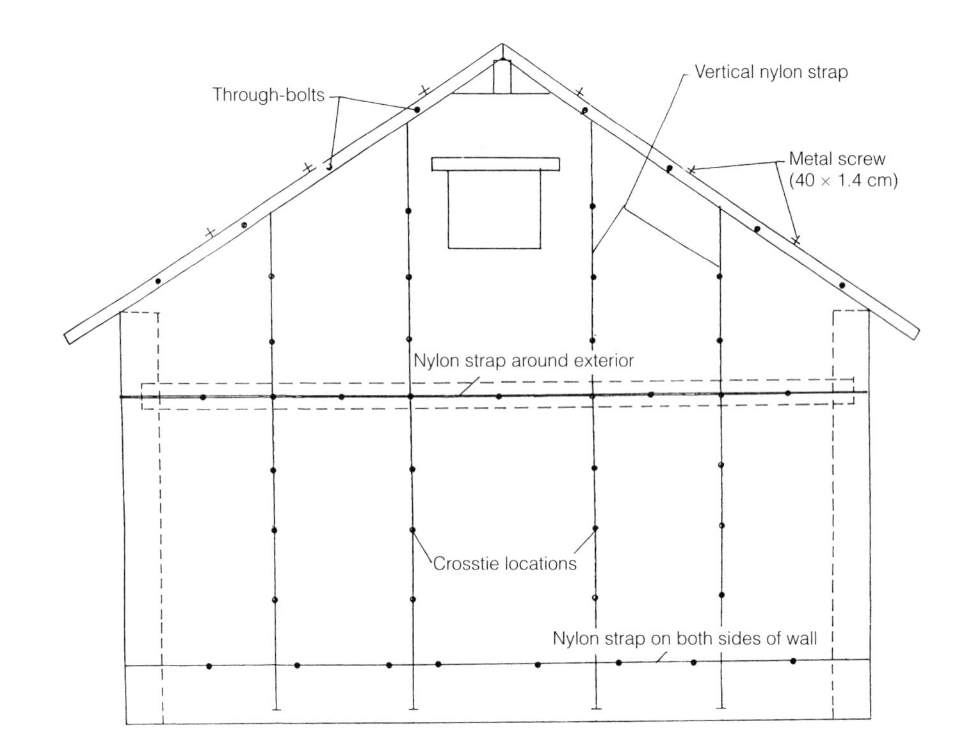

Figure 7.7
Model 11 elevation of east wall.

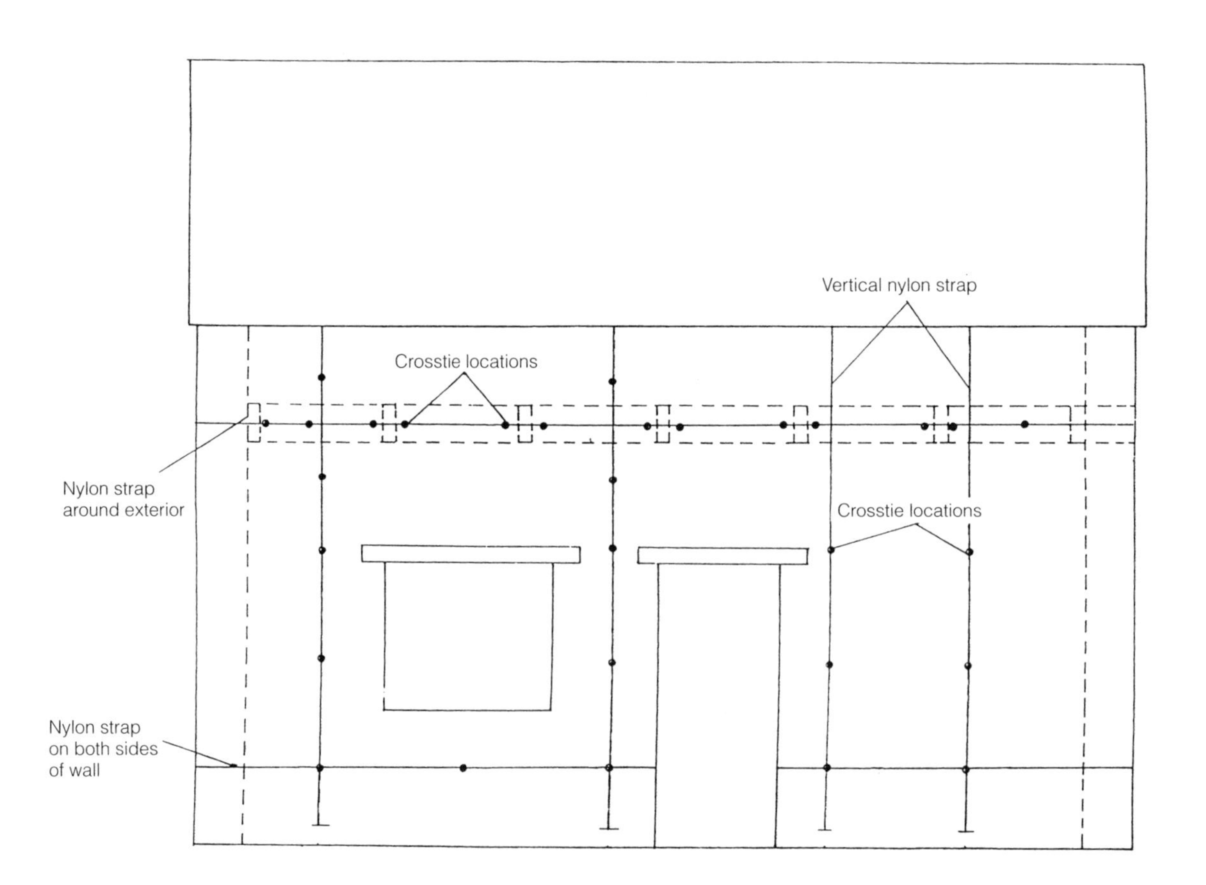

Figure 7.8
Model 11 elevation of north wall.

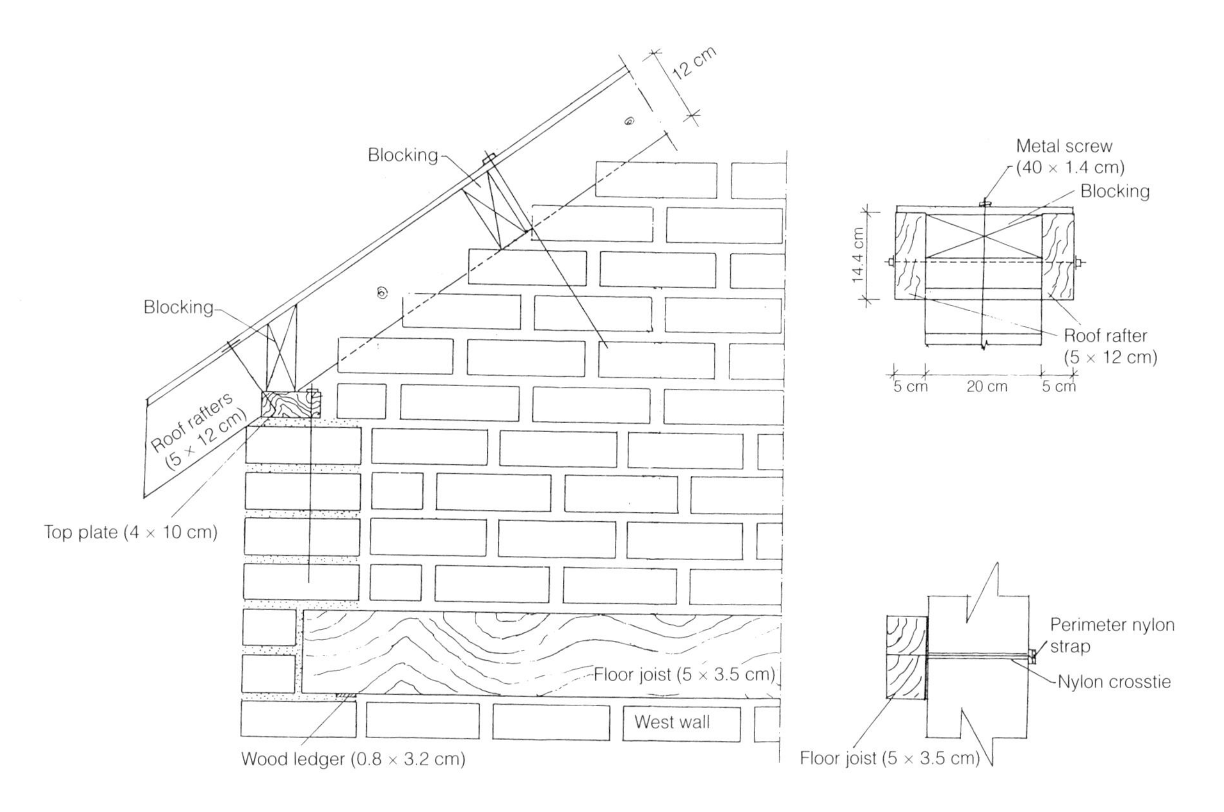

Figure 7.9
Model 11 details of roof and joist attachment.

Partial wood diaphragms were added to the attic floor (fig. 7.10). The width of the diaphragm on the attic floor was approximately 20 cm (7.8 in.) and was equivalent to the spacing between the floor joists. Additional straps were added to the attic diaphragm for continuity. Wood blockings were inserted between the floor joists.

Table 7.2
Measured strength of nylon straps (modulus of elasticity: $E = 5300$ kg cm^{-2})

Tensile force P (kg)	Tensile strength σ (kg cm^{-2})	Strain $\Delta 1/1$ (%)
0	0	0
3	4.2	1
13	18.2	1.3
50	70	2.1
100	140	3.1
150	210	4.2
200	280	5.3
250	350	6.3
300	420	7.2
350	490	8.0
400	560	8.6
450	630	9.1
500	700	9.6
600	840	10.0
650	910	10.7
700	980	11.0
705	987	breaking

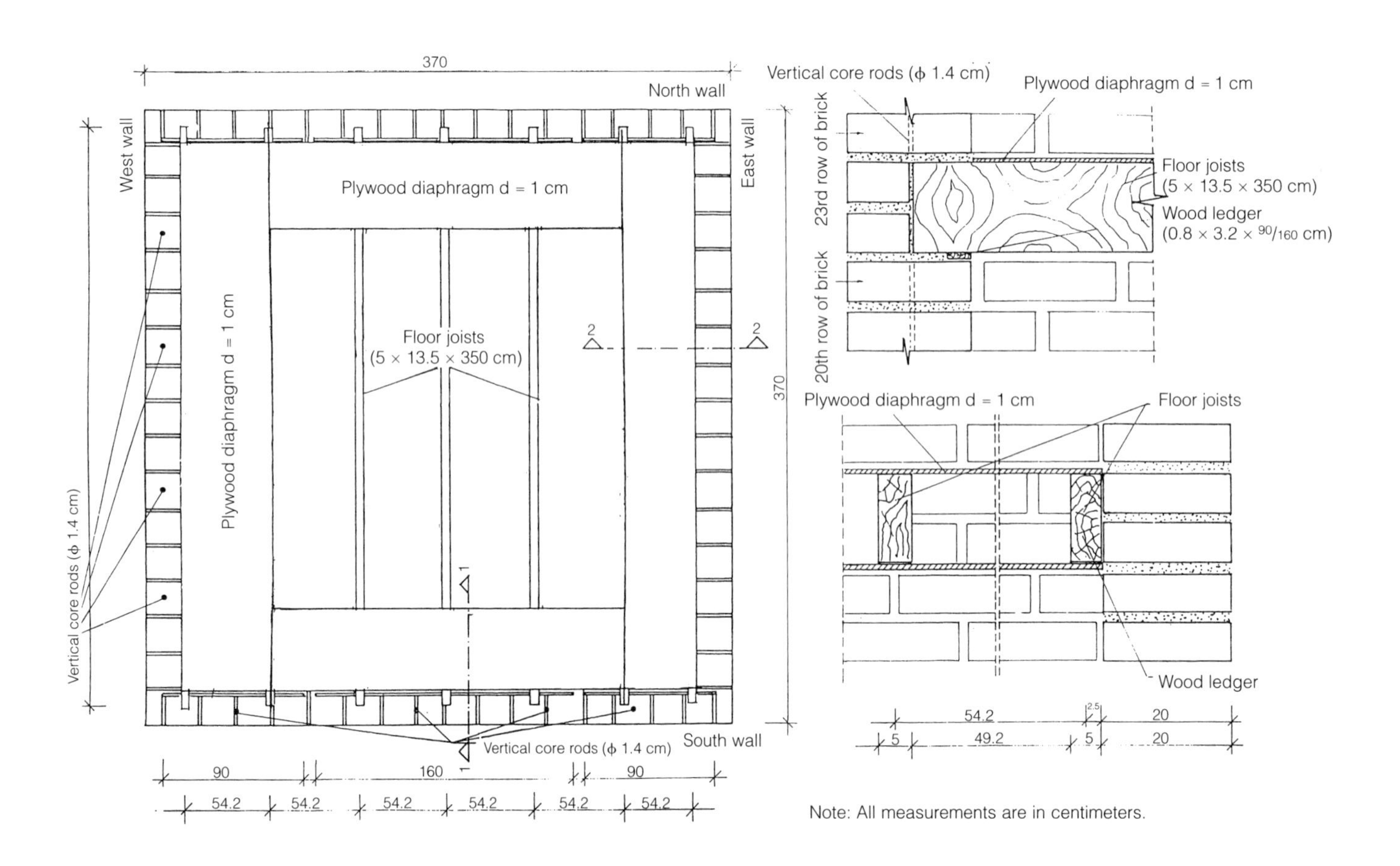

Figure 7.10
Model 11 plan at the level of plywood floor diaphragm.

On the load-bearing walls, the bearing plates on the tops of the walls were discontinuous; that is, they were cut into four sections. These plates were attached to the walls with 8 mm (0.31 in.) diameter steel screws anchored with epoxy grout. The roof rafters were anchored with nails to the bearing plates, and blocking was placed between each of the roof rafters (fig. 7.11).

The floor joists were anchored to the walls with small diameter nylon cable ties. The ties went through holes drilled through the center of the floor joists (fig. 7.9), through the adobe wall on either side of the joists, and were attached to the horizontal straps on the exterior face of the wall.

On the non-load-bearing end walls, the roof rafters were placed directly on both sides of the wall and were tied together with through-bolts (fig. 7.12). The partial roof diaphragm was attached to the tops of the roof rafters. Steel screws 40 cm (9.2 in.) long extended through the roof diaphragm into the wall and were grouted in place with epoxy resin. Unlike the unretrofitted model, plywood diaphragms with a total width of 40 cm (9.2 in.) (at the eaves) and a width of 30 cm (7.6 in.) (at the corners) were applied to the roof. These were connected to the rafters and the top plate by nails.

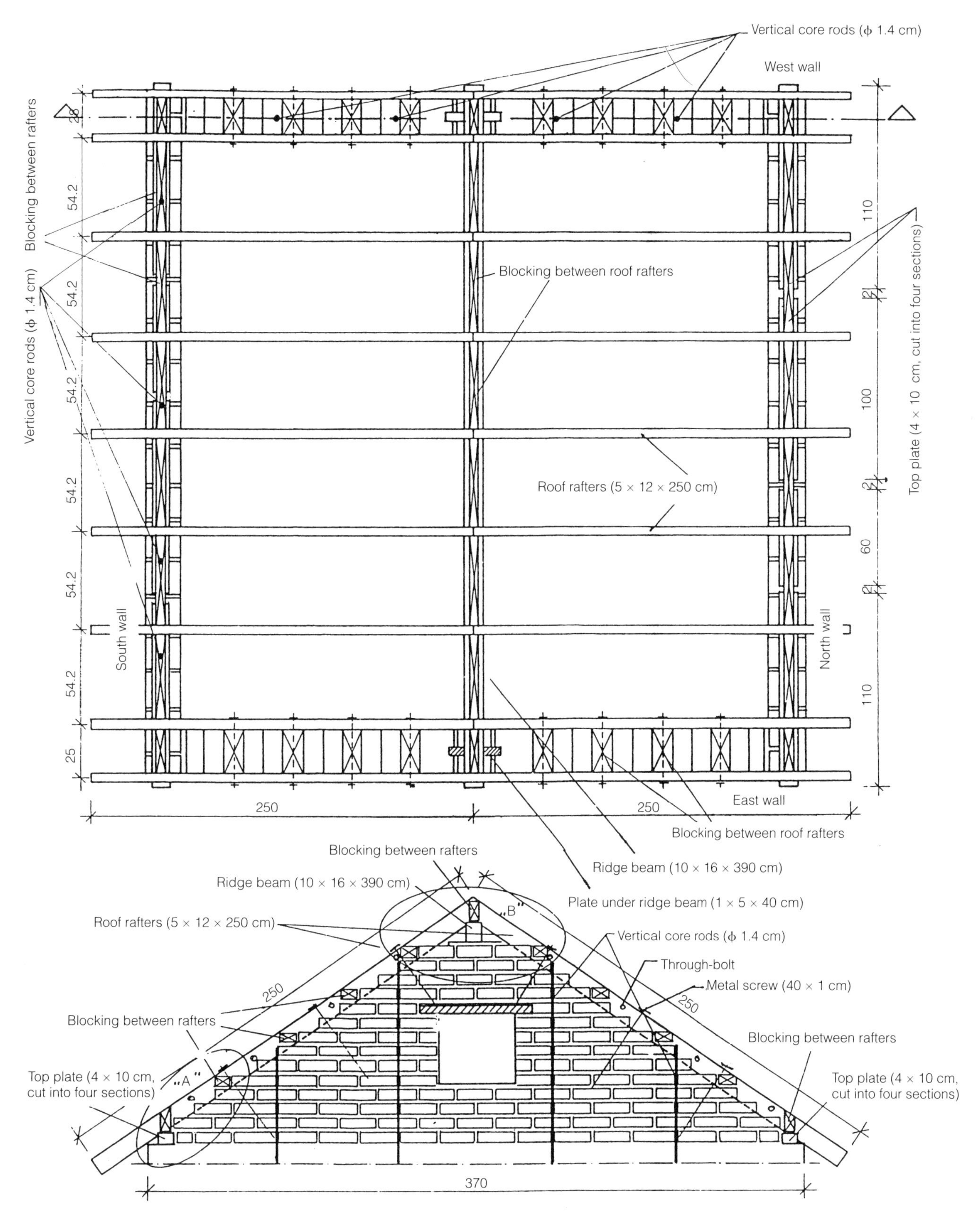

Figure 7.11
Model 11 roof framing plan.

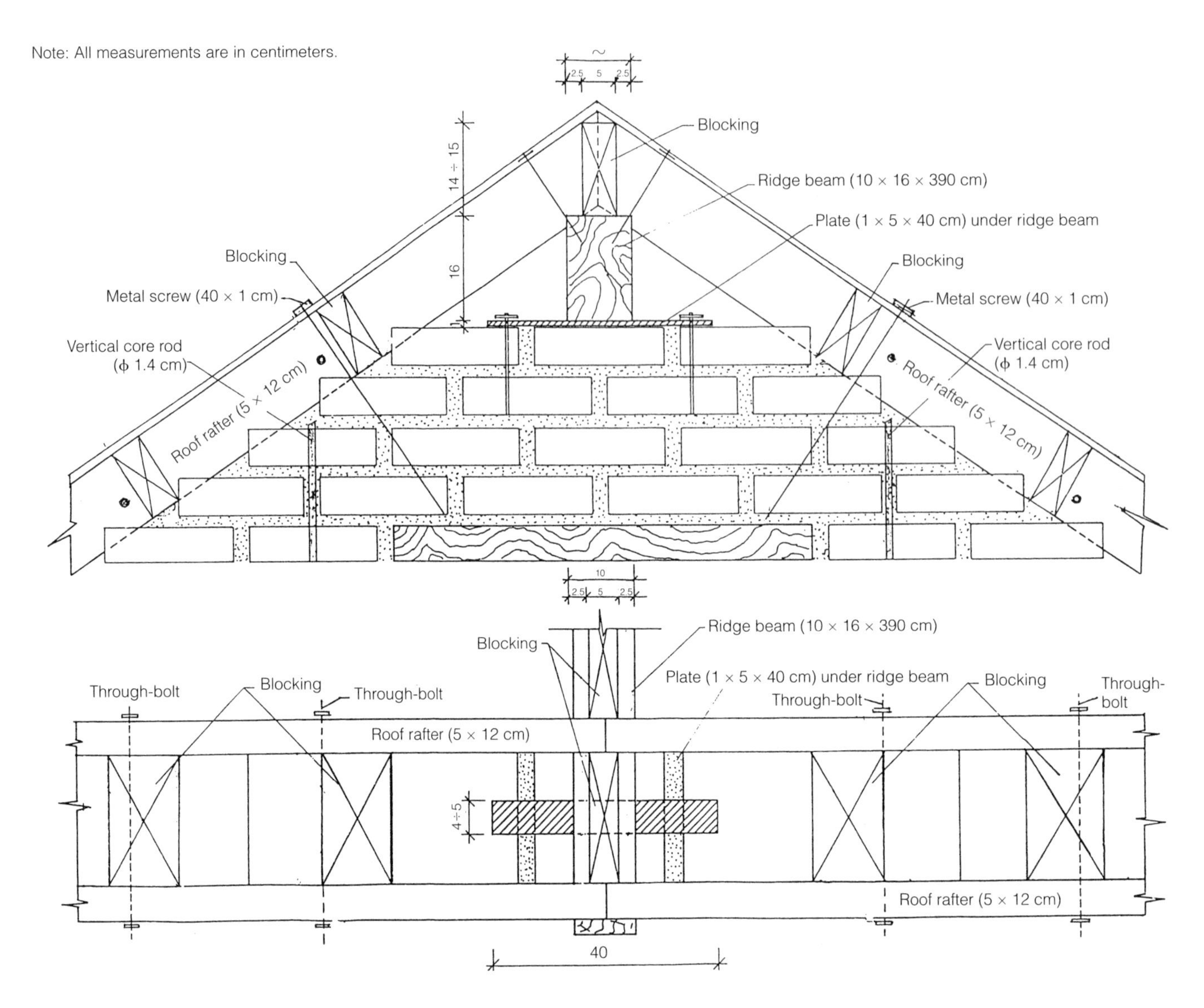

Figure 7.12
Model 11 detail at ridge.

Chapter 8

Test Results and Data Analysis for Models 10 and 11

A description of the behavior of the model buildings during the seismic simulation tests discussed here can be organized by recognizing three stages of response: elastic response, crack initiation and progression, and severe damage and collapse.

During the first stage, the building behaves elastically before cracks affect the dynamic performance. The dynamics of the buildings can be understood by studying the time histories that were recorded by accelerometers and displacement transducers mounted on the building. Frequency characteristics are studied by using those two histories and fast Fourier transforms of the acceleration records. Because the models are stiff, they appear to move in unison with the shaking table. The dynamic response of the model, described by the motion of the model relative to the shaking table, cannot be detected visually because the amplitude of the dynamic motions is small and the frequency of the motions is high.

The second stage is characterized by the initiation and progression of crack damage. The first changes in dynamic motions could be seen only in the acceleration time histories. Cracks then began to develop and continued to develop during the next several tests. The principal crack pattern was completed when the cracked wall elements rocked and rotated around the existing cracks. During the tests on the two models, the cracks in the out-of-plane, gable-end walls were fully developed, but the cracks in the in-plane walls were not yet fully established.

The third stage occurred after the major cracks had fully formed and the stability of the walls was in question. For thin walls, cracks and collapse may occur at the same level of ground excitation, whereas in thicker-walled buildings, such as these models, the tests indicated that cracks develop at levels of excitation well before the stability of the walls is in question. During severe ground shaking, collapse may occur; or, for the retrofitted walls, the retrofit measures become fully engaged, and the instability of the walls may be prevented only by the actions of the retrofit elements.

Model 10—Unretrofitted

Model 10 was an unretrofitted control model. The roof system had minimal, widely spaced roof sheathing. Nevertheless, during large displacements of the gable-end walls, the roof system still had enough stiffness to affect the out-of-plane performance of those walls. The mode of failure was a simple overturning of the wall and an out-of-plane collapse initiated near the mid-height of the wall.

Elastic response

The dynamic response of the undamaged control building was largely as expected. The time histories of the wall and table accelerometers are shown in figure 8.1. By comparing the peak values for each plot, it can be seen that the peak acceleration at the top of the west wall (0.332 g)

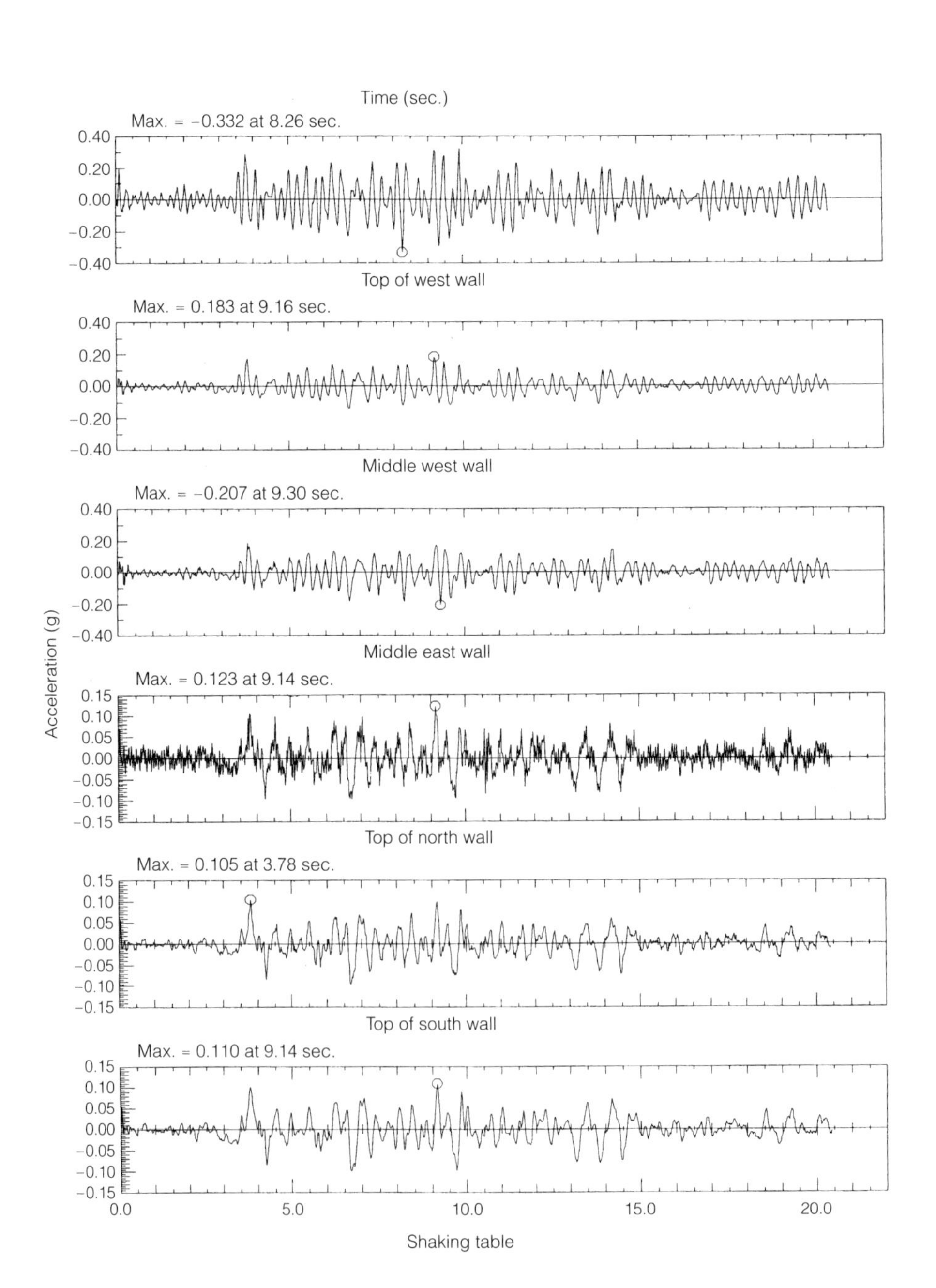

Figure 8.1
Acceleration measurements for model 10, test II. The peak acceleration magnitude and the time at which the peak occurred are indicated by a circle around the maximum peak on the time-history plot. The value on each plot is shown toward the left side just above the top border of each time history.

was approximately three times the peak acceleration of the shaking table, which was 0.110 g. Since the tops of the east and west gable-end walls were coupled by the ridge beam, the acceleration at the top of the east gable-end wall was not measured. At the mid-height of the west and east walls, the peak accelerations were 0.183 g and 0.207 g, respectively, indicating that the accelerations at this location were nearly twice the base acceleration for the out-of-plane walls. Even though the anchorage at this location was not strong, the tops of these walls probably moved in unison during low-level tests.

In contrast, the peak acceleration magnitudes of the in-plane (north and south) walls were only 0.123 g and 0.105 g, respectively. These values indicate that the motion of the in-plane walls was dominated by rigid-body motion; that is, the dynamic motion of these walls closely followed that of the base motion.

Transfer functions show the comparison of the frequency characteristics of one signal relative to a second signal. By examining a transfer function, one can determine the frequency ranges at which the amplification of motion occurs. The transfer functions for the top of the west gable-end wall, the middle of the west gable-end wall, and the middle of the east gable-end wall were determined relative to the table acceleration. The frequency plots indicate that the principal mode was between 8.0 and 8.5 hertz. This value is slightly lower than the 9.8 hertz measured during the forced vibration tests (Gavrilovic et al. 1996). The slightly lower frequency was probably due to the interaction between the shaking table and the heavy adobe structure when the table was in the raised position during the dynamic tests. The forced vibration frequencies were determined when the table was in the lowered position, which resulted in a stiffer foundation system and less structure-foundation interaction.

Displacement time histories were not measured during the low-level tests when the building was undamaged because the expected displacements were too small; these data become important only during high-level tests after cracks have developed.

Damage progression

Crack damage began during test III, when very small cracks occurred in both east and west walls. In the west wall, the crack was horizontal at the height of the north and south walls. In the east wall, the crack was vertical and extended about halfway to the foundation. A small crack appeared in the upper left corner of the north wall caused by the out-of-plane motions of the east wall, which created tensile stresses in the north wall as it began to pull away. The dynamic motions of the upper sections of the walls were slightly discernible during test III and became much easier to see during tests IV and V, when cracking became more significant.

The crack patterns after test V are illustrated in figure 8.2. The cracks in the gable-end walls show the variety and types of cracks that can occur during out-of-plane motions of walls. In the east wall, the vertical crack below the window developed first, followed by the horizontal and diagonal cracks. In the west wall, the horizontal crack developed

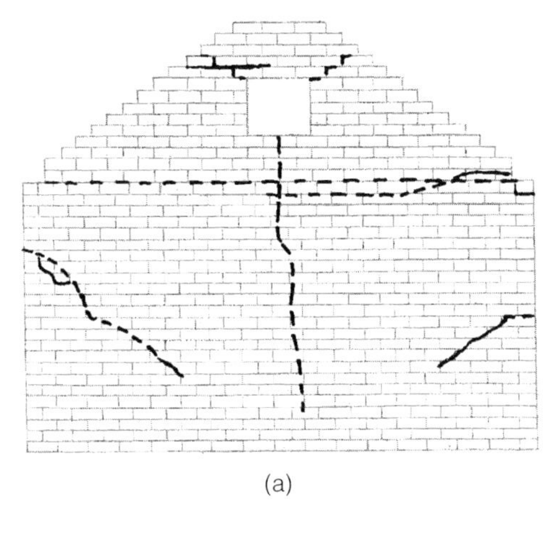

(a)

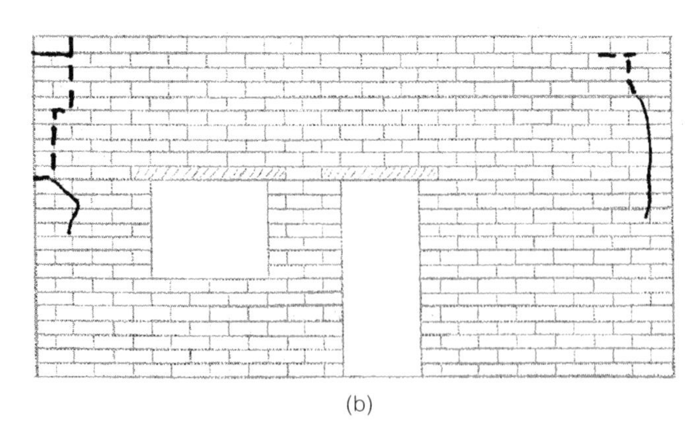

(b)

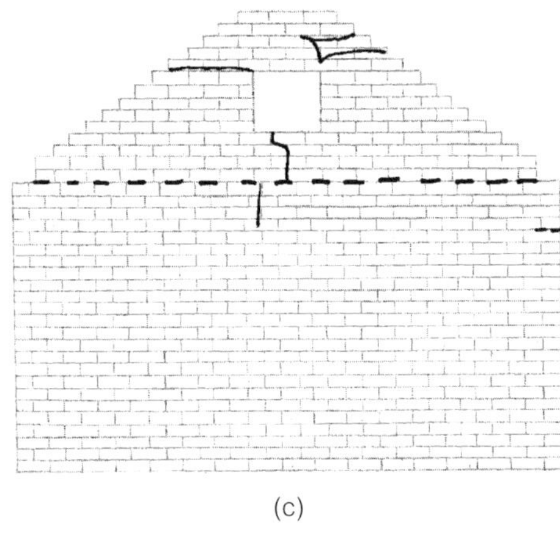

(c)

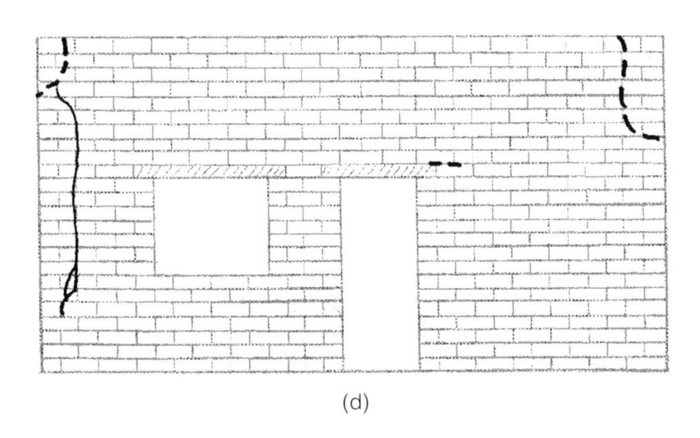

(d)

Figure 8.2
Diagram of model 10, test V, crack patterns on (a) east wall; (b) north wall; (c) west wall; and (d) south wall.

Cracks during this test ————————
Preexisting cracks - - - - - - - - - - - - - - -

first and appeared to have relieved much of the higher tensile stresses in the wall, since there were no diagonal cracks in the lower section and only a very small vertical crack below the horizontal crack. Note that the horizontal cracks in both the east and west walls were located at the tops of the north and south walls. The floor line was several courses lower. Cracks developed at the upper corners of both window openings, as typically occurs at openings in walls. The dynamic motions were apparent during both tests IV and V, with the roof system rocking back and forth significantly. During test V, spalling of adobe along the major lines became more apparent, particularly along the horizontal cracks. The peak displacements at the tops of the gable-end walls during tests IV and V were 7.6 and 9.9 cm (3.0 and 3.9 in.), respectively.

The north and south walls were in plane; that is, the direction of the table motions was parallel to the plane of these walls. The only cracks that occurred during these moderate-level tests were at the ends of the walls, where the tensile stresses were high as a result of the gable-end walls pulling away from the in-plane walls.

Performance and collapse during severe ground motions

Documentation of the progression of damage to the walls of model 10 during tests VI–VIII are shown in figures 8.3–8.7. The stability of the model building was challenged during each of these tests. It is unclear

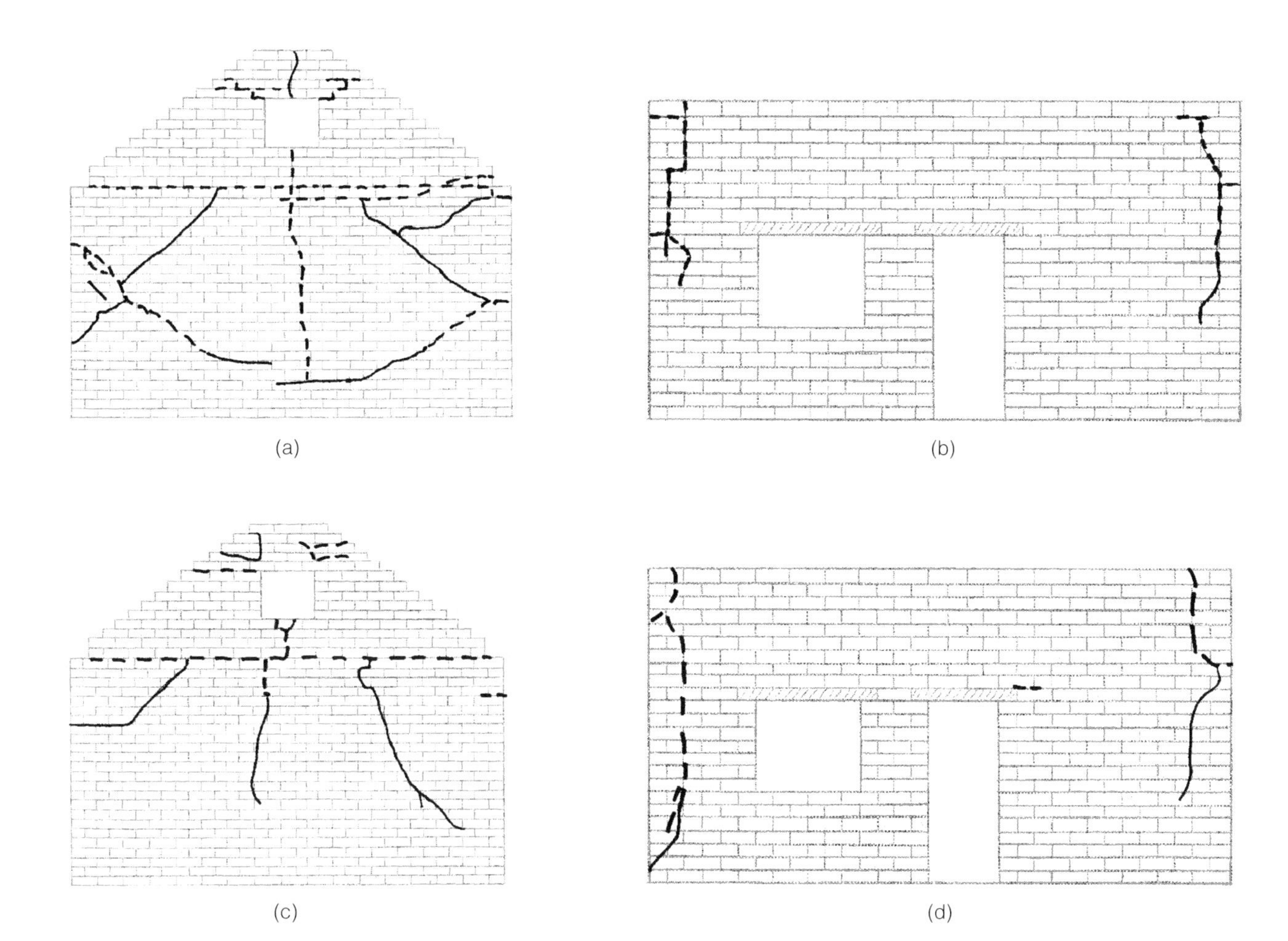

Figure 8.3
Diagram of model 10, test VI, crack patterns on (a) east wall; (b) north wall; (c) west wall; and (d) south wall.

Cracks during this test ————
Preexisting cracks - - - - - - -

to what extent the roof system restrained motions of the out-of-plane, gable-end walls, but the system had a significant effect on the stability of the walls based on the failure mode of the east wall during test VIII. The first failure mode of a wall with little or no restraint at the top is simple overturning. Rather than overturning, however, the east gable-end wall collapsed due to outward motion of the wall sections starting at the upper horizontal crack at about two-thirds the height of the wall.

During test VI (figs. 8.3 and 8.4), the out-of-plane motions of the gable-end walls were significant, with peak displacements of nearly 16.5 cm (6.5 in.). Cracks developed more fully during this test and completed the principal crack patterns for the out-of-plane walls. The patterns that occurred in the east and west walls were different due to the variability in the adobe masonry and the asymmetry of the ground motions. Spalling continued along the major cracks. Without the restraint of the roof system, the gable-end walls would have had a tendency toward instability. Minimal restraint was required to significantly affect the out-of-plane dynamics of thick walls.

During test VII (figs. 8.5 and 8.6), the out-of-plane motions were very great and caused cracking and popping noises in the wood framing of the roof. The out-of-plane walls would very likely have overturned had it not been for the restraint provided by the roof framing system. Spalling continued along the major cracks. Diagonal cracks

(a) (b) (c) (d) (e)

Figure 8.4
Photographs of model 10, test VI, crack patterns on (a) east wall; (b) north wall; (c) west wall; (d) south wall, west side; and (e) south wall, east side.

developed during test VII in the north wall (fig. 8.5b). This was the first test during which typical diagonal in-plane cracks developed; there were no similar in-plane cracks in the south wall. The out-of-plane displacement of the west gable-end wall was 23.6 cm (9.3 in.) during test VII.

The in-plane displacements of the north and south walls were not accurately measured during test VII. Vertical cracks opened near the west end of both the north and south walls, allowing out-of-plane motion of the entire west wall. The opening of the vertical crack in the south wall was visually larger than the vertical crack in the north wall (fig. 8.6d). The peak displacement on the south end was 7.4 cm (2.9 in.), more than three times the peak displacement at the north end.

During test VIII (fig. 8.7), the east wall collapsed from the center of the wall rather than overturning due to restraint by the roof system. Test VIII lasted only about 3 seconds because the programmed table accelerations exceeded the capacity of the earthquake simulator. Due to the short length of the test, little additional damage occurred in the other walls of the building. The overturning failure mode occurs by rotation about the lowest horizontal crack, and the maximum displacement occurs at the top of the wall (fig. 8.8a). This type of collapse is referred to as "mid-height, out-of-plane collapse" (fig. 8.8b).

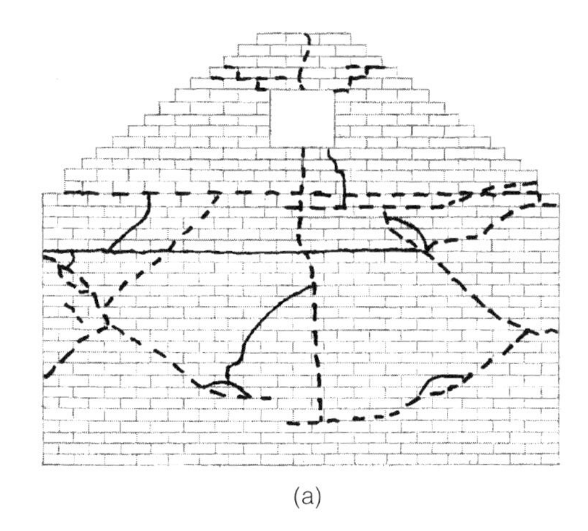

(a)

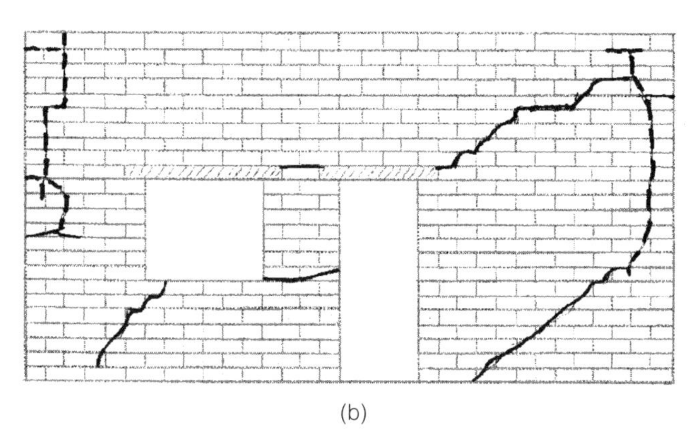

(b)

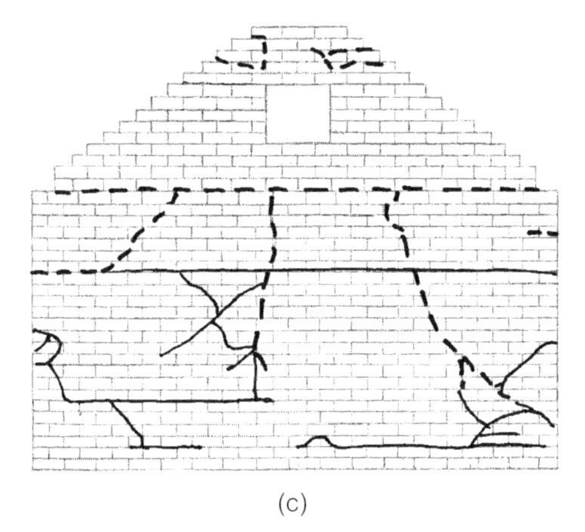

(c)

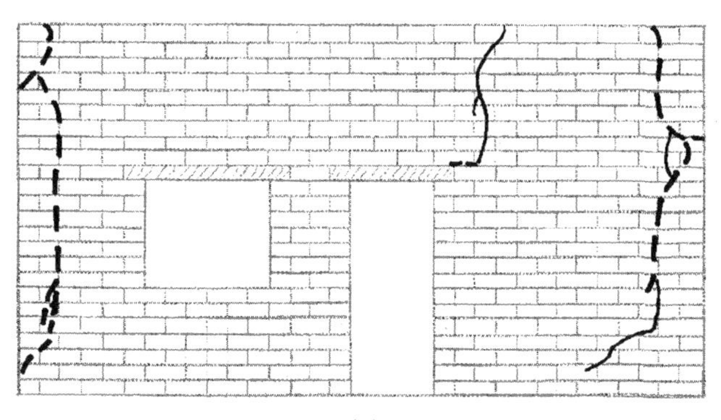

(d)

Figure 8.5
Diagram of model 10, test VII, crack patterns on (a) east wall; (b) north wall; (c) west wall; and (d) south wall.

Cracks during this test ————————

Preexisting cracks - - - - - - - - - - - - - -

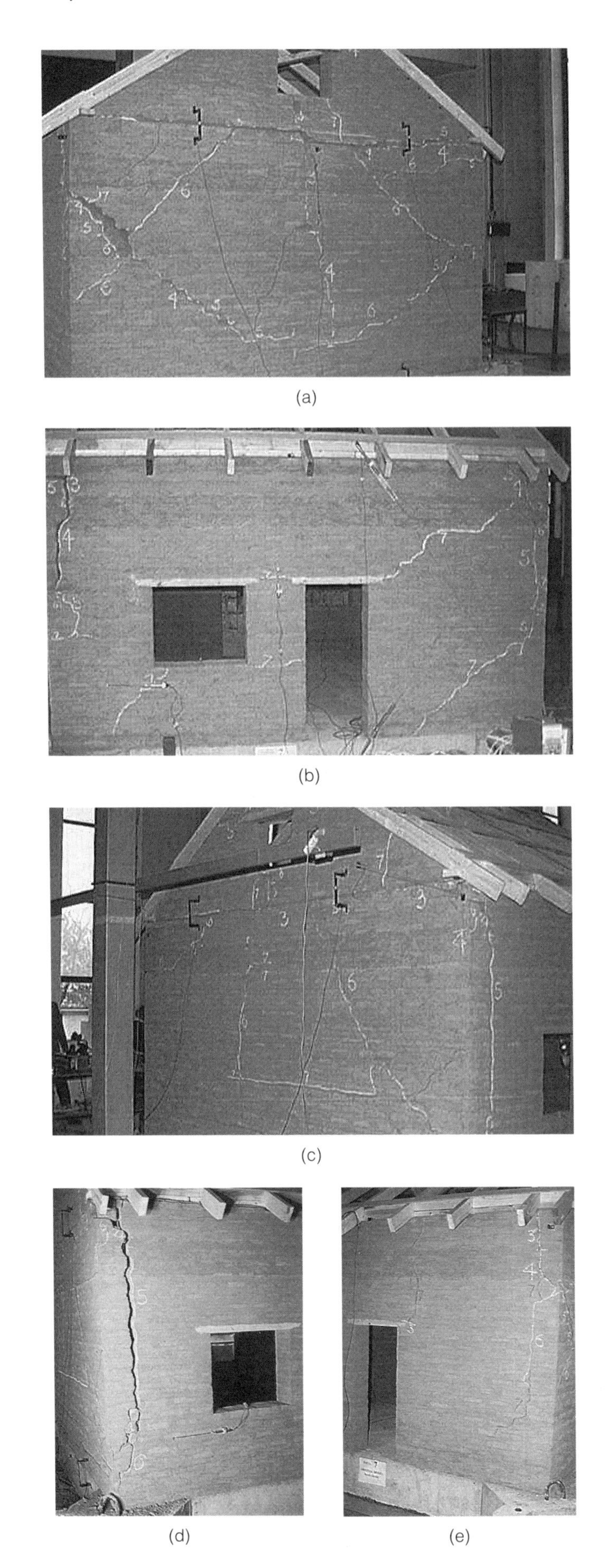

Figure 8.6
Photographs of model 10, test VII, crack patterns on (a) east wall; (b) north wall; (c) west wall, south side; (d) south wall, west side; and (e) south wall, east side.

(a)

(b)

Figure 8.7
Model 10, test VIII, showing (a) east wall collapse; and (b) crack patterns in northwest corner.

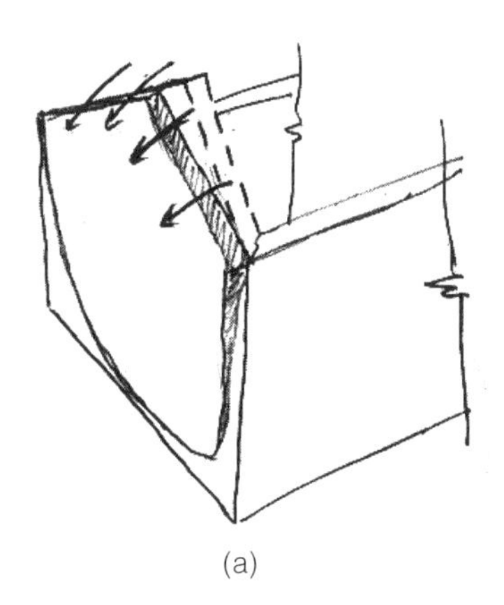

(a)

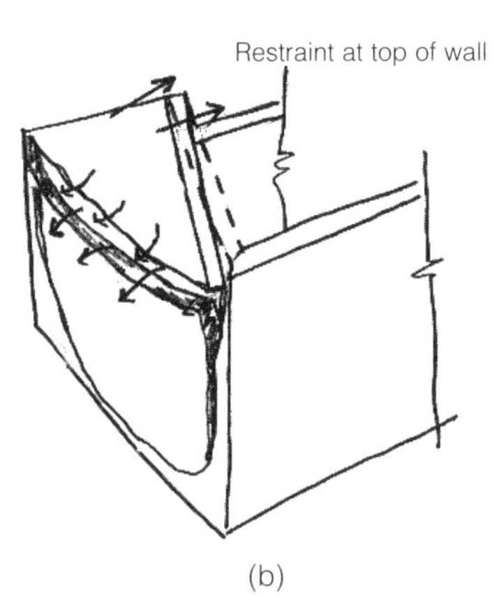

(b)

Figure 8.8
Drawings of out-of-plane modes of failure, illustrating (a) typical overturning failure as the wall rotates about its base; and (b) mid-height, out-of-plane collapse observed in the east wall of model 10. The latter mode of collapse occurred due to the restraint provided by the roof system at the top of the wall.

Model 11—Retrofitted

The performance of the retrofitted model demonstrated the effectiveness of the installed retrofit system. Two different retrofit measures were applied to the two sets of in-plane and out-of-plane walls. The systems were the same except for the difference in the vertical elements: the north and east walls were fitted with vertical straps on both sides of the walls; the south and west walls were fitted with vertical center-core rods.

The elastic performance of the model may have been affected by damage to the retrofitted model that occurred before testing began. The unretrofitted model 10 was constructed on a concrete slab that was already in place on the shaking table. Model 11, however, was constructed on a concrete slab that was then moved to the shaking table. When this slab was bolted to the shaking table, it flexed and produced a deflection that resulted in minor crack damage to the building in the northeast corner.

Elastic response

The elastic response of the model may have been slightly affected by either the preexisting cracks or the installation of the retrofit system. The modal frequencies in the east-west direction as determined from the ambient vibration and forced vibration tests were 11.0 and 11.2 hertz, respectively (Gavrilovic et al. 1996). These values are slightly higher than those of model 10 and may be attributed to the stiffness of the roof and floor diaphragms or to variation from one model to the other. The more important observation is that the elastic behavior was not significantly affected by the preexisting cracks. Nevertheless, the locations of new cracks were clearly predetermined by the existing cracks.

Plots of the dynamic motions of the walls during test II are shown in figure 8.9. The peak acceleration of the top of the out-of-plane west wall was 0.359 g compared to the table acceleration maximum of 0.111 g, an amplification factor of 3.25. The accelerations at the mid-

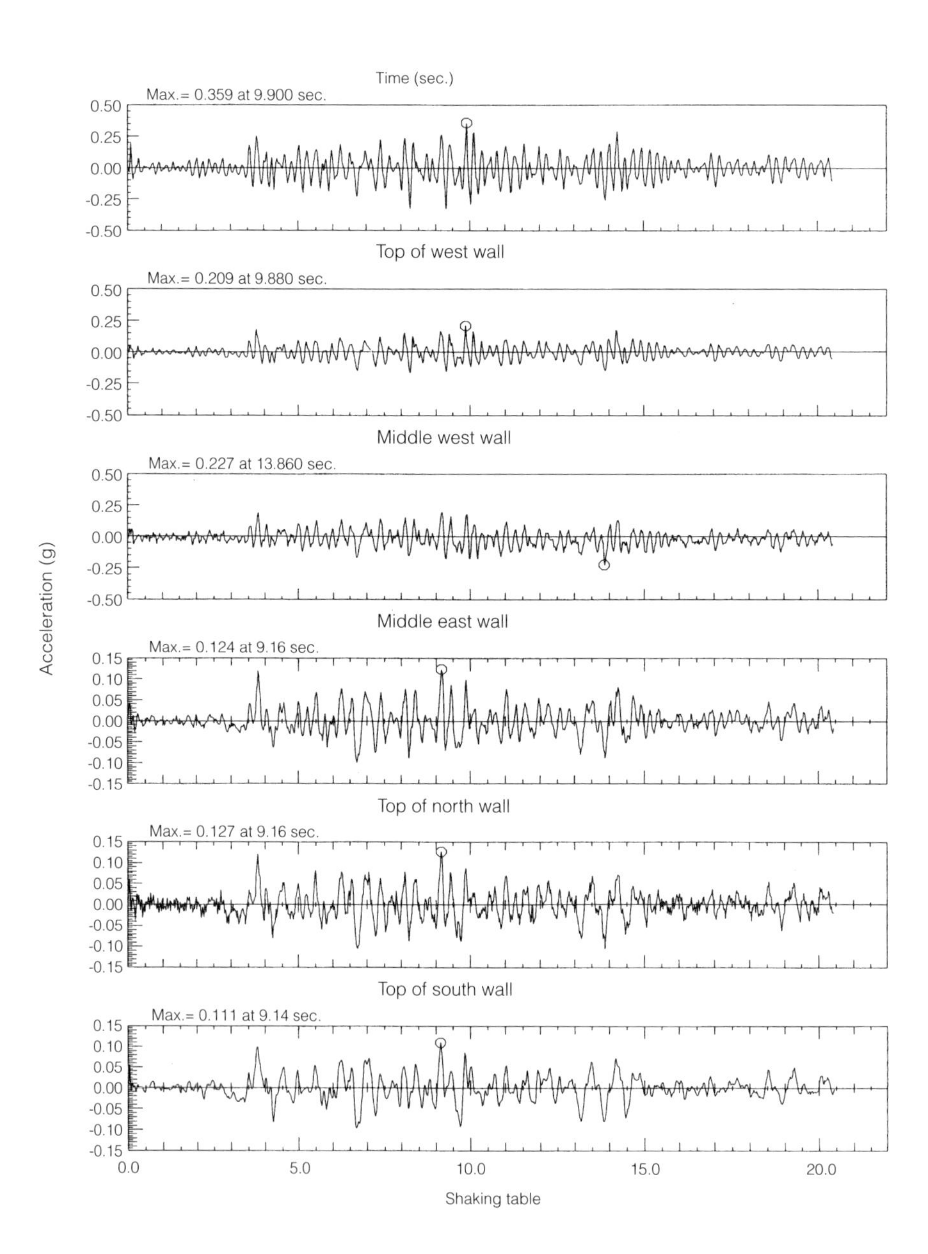

Figure 8.9
Acceleration measurements for model 11, test II. The peak acceleration magnitude and the time at which the peak occurred are indicated by a circle around the maximum peak on the time-history plot. The value on each plot is shown toward the left side just above the top border of each time history.

height of the west and east walls were 0.209 g and 0.227 g, respectively. The accelerations of the north and south (in-plane) walls were 0.124 g and 0.127 g, respectively, only slightly greater than the table acceleration. It is also interesting to note that the peak acceleration of the damaged north wall was the same as that of the undamaged south wall.

The transfer functions for the wall accelerations of model 11 were similar to those of model 10, except the amplification shown for the principal mode of vibration was about 8.5 or 9.0. This value is, again, slightly lower than that measured during the ambient and forced vibration tests. All modal frequencies of model 11 were slightly higher than those of model 10.

Damage progression

New crack damage did not develop until test IV or propagate until test V. The location of the crack damage after test V is shown in figure 8.10. Damage to the two out-of-plane walls was similar in some respects to that in model 10 but different in others, indicating the influence of the retrofit strategies. The vertical straps used on the west walls were not stiff enough to prevent or change the development of crack damage in the wall. The strap system affected only the dynamic performance after cracks had developed. As a result, the crack pattern in the east out-of-plane, gable-end wall consisted of both vertical and horizontal cracks similar to those

Figure 8.10
Diagrams of model 11, test V, crack patterns on (a) east wall; (b) north wall; (c) west wall; and (d) south wall. Earlier cracks are marked with zeros and dashed lines.

Cracks during this test
Preexisting cracks
Straps
Center cores

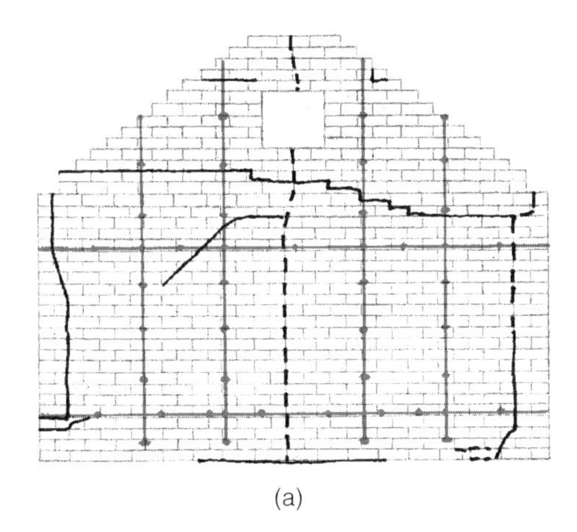

(a)

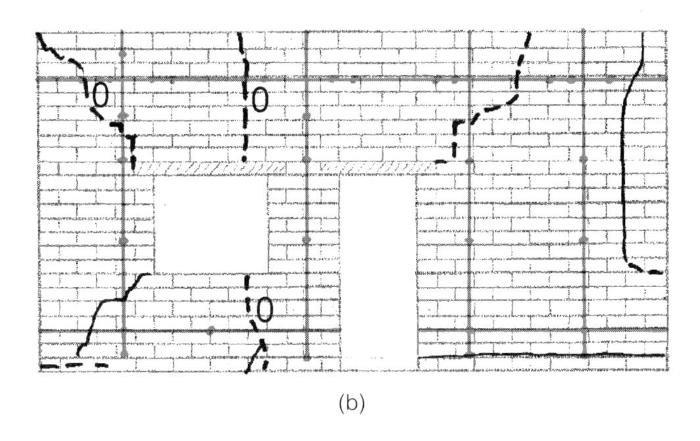

(b)

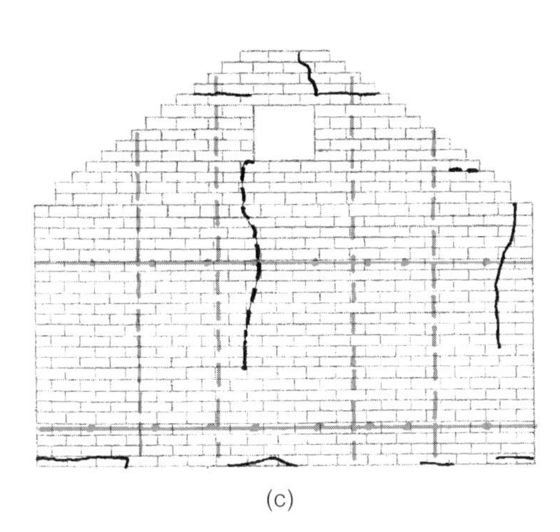

(c)

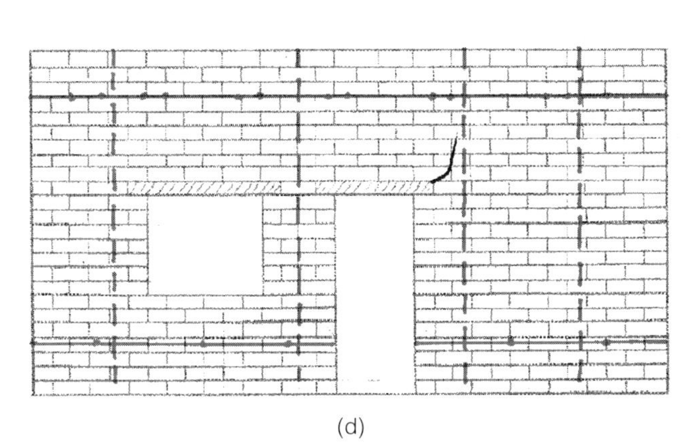

(d)

that occurred in model 10. The horizontal crack at or near the tops of the in-plane walls was, again, similar to the damage in model 10.

Installation of center-core rods in the west wall of model 11 had a clear effect on both the elastic and post-elastic behavior. The center-core rods were anchored into the walls with epoxy grout, which provided sufficient strength and elasticity to allow shear transfer between the adobe wall and the rods. As a result, the development of horizontal cracks was delayed until test VII. The vertical reinforcing rods strengthened the adobe wall against flexure about the horizontal axis.

Since there were no horizontal rods, vertical cracks developed at roughly the same time as the cracks in the east wall. Restraint against deflection about the vertical axis was provided only by anchorage to the roof and floor diaphragms and the horizontal strap. These restraints were not sufficient to prevent vertical cracks from occurring, however.

The in-plane north wall with vertical and horizontal straps developed additional cracks during both tests IV and V (fig. 8.10b). These cracks occurred earlier than in the in-plane walls of model 10 for two reasons: (1) some cracks existed before the testing began, and (2) the retrofit system resulted in higher forces being transferred to the in-plane from the out-of-plane walls. More than likely, both factors contributed to greater cracking in the north wall of model 11 compared to the north wall of the unretrofitted model 10.

The south wall with vertical center-core rods suffered little damage. A minor crack began at the upper right corner of the in-plane south wall but appeared to stop when the crack reached the epoxy-anchored, center-core rods.

In viewing the dynamic tests, the cracks that developed during tests IV and V were relatively minor in terms of the amount of displacement that occurred in the out-of-plane walls. In the photographs of figure 8.11, it is difficult to discern the locations of the cracks because they were highlighted with chalk only, and no spalling had occurred around the cracks. Not until test VI did larger displacements develop and spalling become significant.

Performance during severe ground motions

During the strong ground motions of tests VI–VIII, all of the retrofit measures on the building became significant with regard to the performance of model 11. The walls with vertical straps, both in-plane and out-of-plane, sustained much higher levels of damage than the walls with the vertical center cores. The drawings and photographs of model 11 after test VII are shown in figures 8.12 and 8.13.

During test VI, the east wall began to sustain significant displacements, and spalling occurred along crack lines, particularly along horizontal cracks. The most significant additional cracks were in the east wall. Despite the spalling of material, the building did not suffer any measurable offsets at the cracks.

Test VII was informative because of the significant difference in the behavior of the two sets of walls. The north and east walls with vertical straps suffered large displacements, spalling, and significant off-

Figure 8.11
Photographs of model 11, test V, crack patterns on (a) east wall; (b) north wall; (c) west wall; (d) west wall gable; (e) south wall, west side; and (f) south wall, east side.

sets, whereas the south and west walls with center-core rods sustained relatively minor damage. Cracks continued to develop in the south and west walls, but these were relatively minor, and there were no offsets.

The extent of damage to the north and east walls contrasts sharply with the low level of damage in the south and west walls. Spalling continued, with large cracks opening on the east and north walls and permanent offsets at several cracks. The dynamic motions of the east gable-end wall were very great, causing relatively large loads to develop in the vertical and horizontal straps.

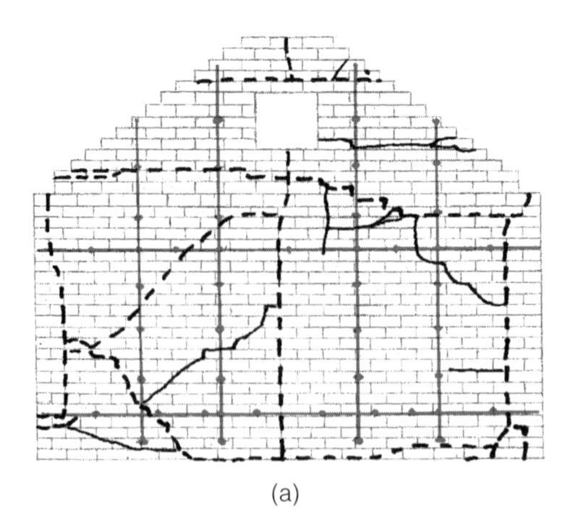

(a)

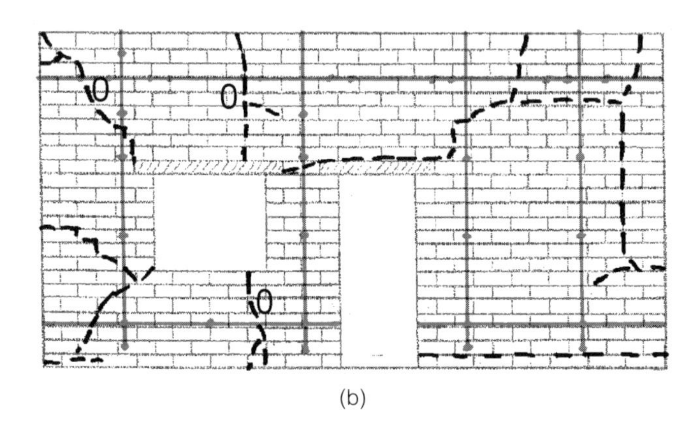

(b)

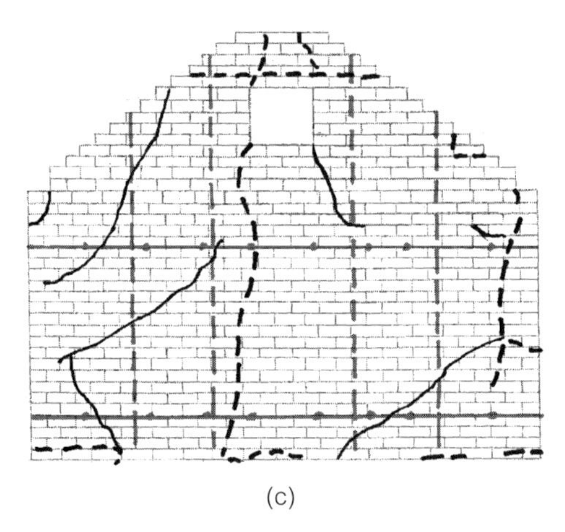

(c)

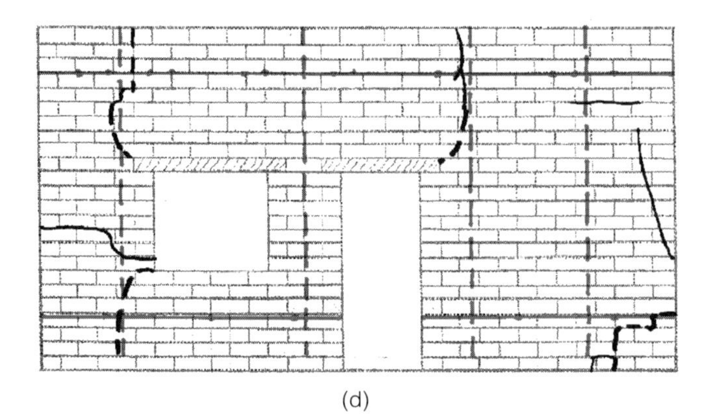

(d)

Figure 8.12

Diagram of model 11, test VII, crack patterns on (a) east wall; (b) north wall; (c) west wall; and (d) south wall.

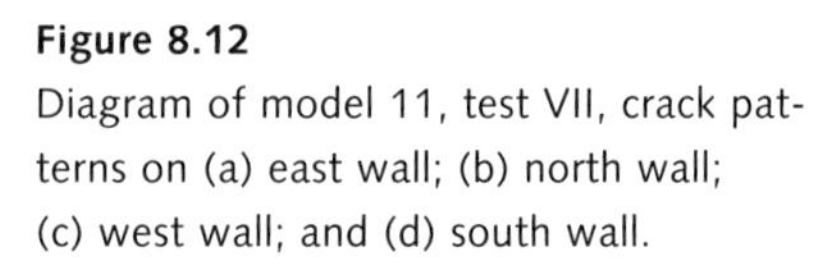

Figure 8.13
Photographs of model 11, test VII, crack patterns on (a) east wall; (b) north wall; (c) west wall gable; (d) west wall, south side; (e) south wall, west side; and (f) south wall, east side.

The most significant damage may have been the displacement of the block at the east end of the north wall (fig. 8.13b). The section of the wall defined by the diagonal cracks at the upper and lower left corners of the window created a cracked wall section along which the permanent displacements continued to grow as the tests proceeded, beginning with test VII. The crack at the top of this panel developed a permanent displacement of approximately 3.8 cm (1.5 in.) during this test, and significant loads were applied to both the upper and lower horizontal straps. The load on the upper strap was great enough to result in a secondary horizontal crack that developed just below this strap. This section of wall was difficult to restrain because it continued to move downward and out along the lower diagonal crack. Gravity and dynamic motions continually moved this section outward. In this type of damage, a diagonal crack at the end of a wall section creates a condition in which damage progressively worsens during continued earthquake motions and may eventually lead to catastrophic failure.

The shaking-table motions for test VIII* on model 11 were slightly smaller than those for the same test on model 10. During the latter test, the programmed table motions exceeded the acceleration capacity of the shaking table, and the system shut down after the first few seconds of the test sequence. The acceleration for test VIII* on model 11 was decreased to try to prevent overloading; this was not successful, and the shaking table stopped after a few seconds into the test sequence.

Several other attempts were made using a filtered displacement record, but since the table motions were dominated mostly by large accelerations, the table capacity (determined by the force capacity of the system) was exceeded and the system shut down each time.

Large displacements were obtained during test VIII* by decreasing the time-scaling factor from 2 to 1.5. Because the model was 1:2 scale and neglected gravity loads, the correct test acceleration was obtained by multiplying the original accelerations by a factor of 2. Once the model was severely cracked, however, the development of large displacements was the most characteristic effect of the ground motions. Test VIII* had the same table displacements as the previous test VIII for model 10, but the duration of the test was slightly longer. All accelerations were 50% greater, rather than 100% greater, than the original record. Since the force on the drive system for the shaking table is a function of peak acceleration times mass, this decrease was sufficient to prevent overloading of the shaking-table control system. Figure 8.14 shows model 11 after test VIII*.

The east gable-end wall suffered significant additional damage during test VIII*. A small block just below the window was dislodged (see fig. 8.14a), and the offsets at the cracks in the center were nearly 7.6 cm (3 in.). Although large dynamic displacements occurred during the test, the wall never appeared to have been close to collapse.

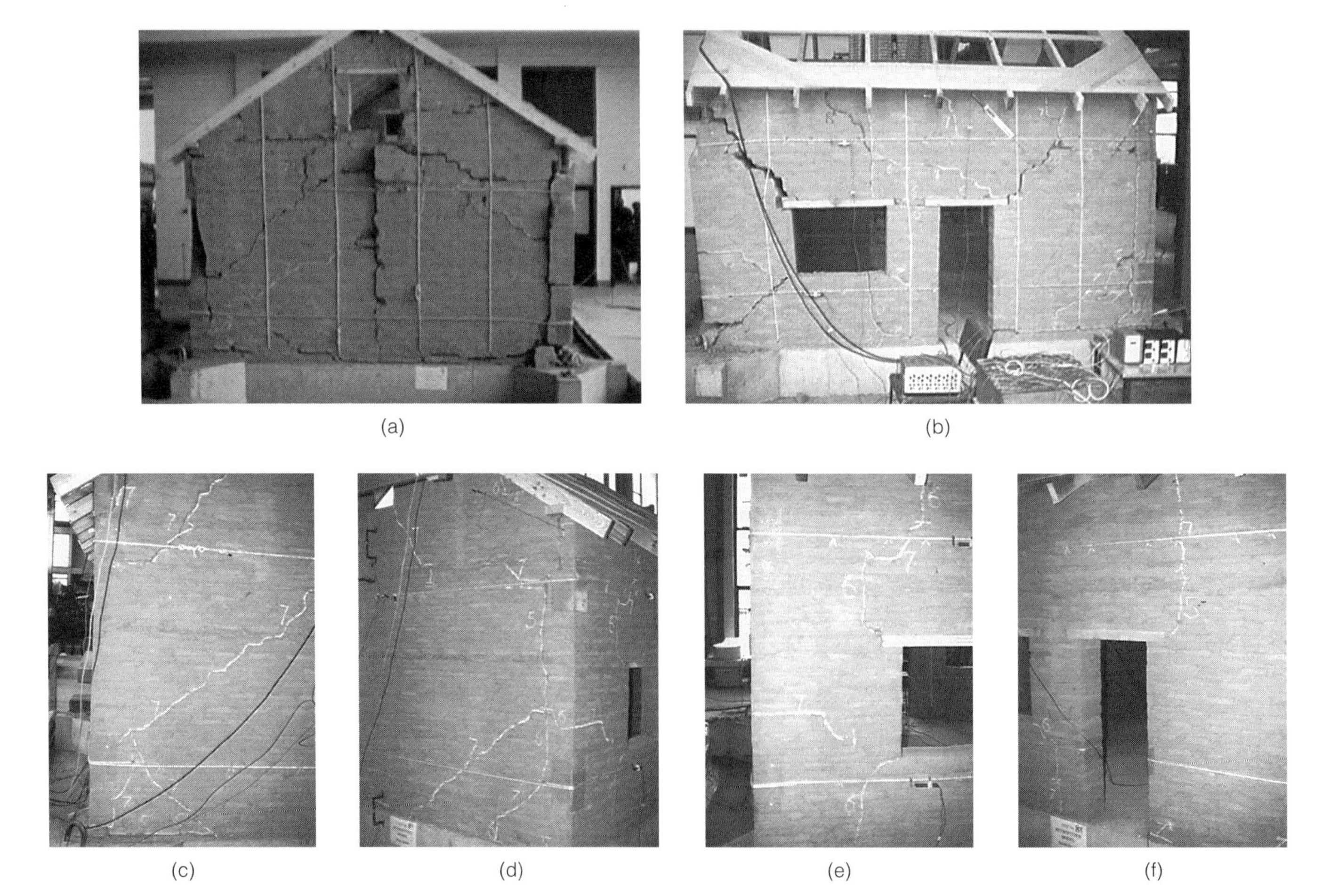

(a) (b)

(c) (d) (e) (f)

Figure 8.14
Model 11, test VIII*, crack patterns on (a) east wall; (b) north wall; (c) west wall, south side; (d) west wall, north side; (e) south wall, west side; and (f) south wall, east side.

Damage to the north wall continued to worsen, with additional cracks at the upper right corner of the door opening. The large offset at the cracked wall section on the left (east) end of the wall became slightly greater, but the tension in the upper and lower straps was sufficient to prevent considerably larger displacements from occurring at this location.

The west and south walls continued to perform very well. The crack pattern generally remained stable, and no significant spalling occurred. There were no offsets at any of the cracks in these walls, again demonstrating the effectiveness of the center-core rods.

During test VII, the maximum displacements took place at the top of the gable end of the structure and measured 7.9 cm (3.1 in.). The maximum displacement at the attic-floor level was 5.3 cm (2.1 in.). In-plane displacements were measured at the west end of the building with linear potentiometers. As a result, the measurements were undoubtedly affected by crack damage near the west wall. The north wall suffered more damage than the south wall and accordingly showed a large peak displacement of 5.3 cm (2.1 in.), whereas the maximum displacement of the south wall was just less than 2.5 cm (1 in.). Neither of the in-plane walls, particularly the north wall, was displaced as much as these measurements indicated, but the cracked section closest to the measuring device was displaced by this large amount.

Displacement of the top of the gable-end wall during test IV was slightly less than 2.5 cm (1 in.) and increased to nearly 8.5 cm (3.4 in.) during test VIII*. These displacements were considerably smaller than those of model 10.

Loads on Elements of the Retrofit System

Loads on the elements of the retrofit system of model 11 were measured during the test sequences. The elements measured were the horizontal straps, the vertical straps on the east wall, and the ties to the floor framing on both the east and west walls. Once again, all measurements were converted from the model to the prototype domain. To do so, the forces measured in the model must be multiplied by a factor of 4 (the square of the inverse of the scaling factor).

Loading in the horizontal strap was highest on the east wall, where the vertical retrofit element consisted of exterior straps. The west wall was retrofitted with vertical center-core elements, which provided more restraint to the out-of-plane movements of the gable-end wall than that provided by the vertical straps. The extent of damage during higher-level tests and the number of permanent cracks in the east and west walls indicate the extent to which the horizontal straps were functioning. The east wall sustained many more cracks and permanent offsets than the west wall. During the dynamic motion tests, the magnitude of out-of-plane motions was considerably greater in the east wall, resulting in high load levels in the east wall horizontal strap.

The load levels in the horizontal straps on the east wall were highest in tests VII and VIII*. Maximum load was 286 kg (631 lb.) in the prototype domain. Because the upper horizontal strap was continuous around the building, the loads measured in the west wall may have been affected by those in the east wall. However, the strap could have lost some tension around each of two corners before reaching the west wall, and therefore it is not possible to know if the measured loading in the east wall influenced the dynamic motions of the west wall. The east wall loads were probably not affected by the motions of the west wall, given the low displacement levels and the minimal extent of crack damage. The loads in the west wall horizontal strap were less than 25% of those measured in the east wall horizontal strap.

Loads in the vertical straps on the east wall were less than those in the horizontal straps. The loads were highest in tests VII and VIII*, with a maximum value of nearly 55 kg (120 lbs.). The highest load in the vertical straps occurred during test VII, whereas the highest load in the horizontal strap occurred during test VIII*.

Measured loads in the crossties were much higher during test VIII* than in the previous tests for crossties in the east wall. Crossties connected the horizontal strap to the floor diaphragm, and maximum load values were about 172 kg (380 lbs.). The loads were considerably lower in the west wall, only 33 kg (72 lbs.). The difference in values was a clear indication of the level of restraint provided by the straps. The crossties through the east wall were very highly loaded because high levels of restraint were required. Alternatively, the primary strengthening and stabilizing element in the west wall was provided by the center-core reinforcing elements, which resulted in lower loads in the crosstie elements.

The measured loads in the straps and crossties could be used as a preliminary guide for designing the elements of a similar retrofit system. Rough calculations could be made using the geometry of the retrofitted model to determine the levels of loading that would occur. However, one difficulty is that the strap loading is determined by contributions from both the vertical and horizontal components. Further detailed measurements of strap and crosstie loads would be required to develop reliable engineering design data. Nevertheless, the magnitude of these loads was considerably less than would have been predicted by a simple static analysis.

The key to a structural retrofit system is its ductility; it is difficult for any of these connections to fail. The connections are ductile because crossties can yield, bearing areas can dig into the adobe, and straps (in both directions) can also stretch to redistribute the loading.

It would not be appropriate to use the loads developed in this testing program for a retrofit system with nonductile connections. Brittle connections must be designed for much higher load levels. In general, brittle connections are greatly discouraged in adobe retrofit systems. Shear anchors are ductile because the adobe around the anchors will most likely fail. Tension anchors are brittle and cannot be recommended for the design of retrofit systems in adobe buildings.

Analysis and discussion of results

The effectiveness of the retrofit system was evidenced by the improved performance of the walls of the retrofitted model. Both the retrofit system with the vertical straps and the system with the center-core rods provided a great improvement in the ductility of the structural system compared to the unretrofitted model. The performance and collapse of the unretrofitted model demonstrated the fragility and danger of unreinforced adobe construction. Both types of retrofit systems reduced the amount of severe damage and dramatically decreased the risk of collapse. The set of walls retrofitted with vertical center-core rods also delayed the development of cracks and reduced the extent of moderate damage.

Following is a brief comparison of the behavior of the models during the three response stages. (A more detailed comparison is given in Tolles and Ginell 2000.)

Comparison of elastic response

The elastic response of the two model buildings was similar in many respects. The largest factor affecting the elastic response was probably the additional stiffness added to the roof and floor systems of model 11. As a result, the frequency of the principal mode of vibration, as determined by the ambient vibration tests, increased from 9.8 hertz in model 10 to 11.2 hertz in model 11. The preexisting crack damage to model 11 appears to have had no significant effect on the principal mode of vibration.

Amplification of peak acceleration at the top of the gable-end walls (out of plane) was approximately the same for both models. Table acceleration was nearly identical for both models during test II (PGA = 0.11 g), and peak acceleration at the tops of the walls was 0.33 g in model 10 and 0.36 g in model 11. This slight difference (less than 10%) may have been the result of the change in first-mode frequency but, as is more likely, may be simply the random variation between test results. In either case, the difference is not significant.

In general, there was little amplification at the top of the in-plane walls. The peak accelerations measured at the tops of the north and south walls of model 10 were 0.123 g and 0.105 g, respectively. In model 11, the peak accelerations were 0.124 g and 0.127 g for the north and south walls, respectively. The results may have been more consistent in the two in-plane walls due to the effect of the roof and floor diaphragms on tying the building together. An additional high-frequency component shown in model 10 may have been a result of model behavior but more likely was a function of the instrumentation or of how the instrument was mounted.

Comparison of damage progression

The first evidence of damage can be seen by examining degradation in acceleration transfer functions. Figure 8.15 shows plots of the transfer function for model 10 during tests I–IV. There was little change between tests I and II, as indicated by the similar shapes of the transfer functions. Significant degradation began during test III with the loss of the defined

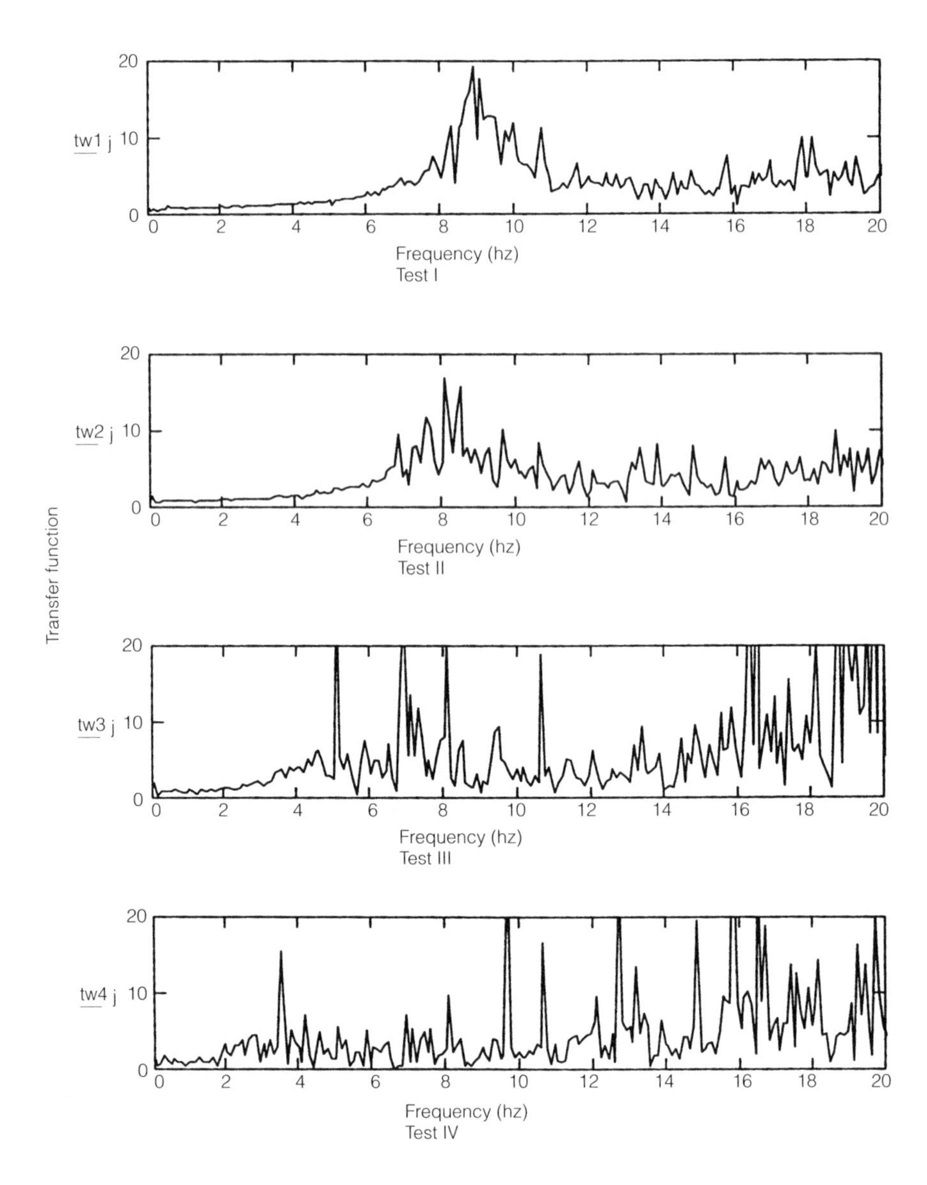

Figure 8.15
Degradation of the frequency response of model 10 during tests I–IV. Plots are the acceleration transfer functions for the top of the west gable-end wall.

peaks in the 8–9 second range, which represents the principal mode of vibration, and with the shortening of the flat part of the spectrum starting at zero and extending to the right. The spectrum of test IV shows complete degradation even in the lowest part of the spectrum.

Comparison of the damage to the out-of-plane and in-plane walls for the two models is shown in figures 8.16 and 8.17, respectively. Drawings of the crack damage that developed during tests V–VII are shown in figures 8.18–8.23.

Crack development of the out-of-plane walls during moderate-level tests was affected by the retrofit system. The stiffness of the roof diaphragm was sufficient to prevent horizontal cracks from occurring in both the east and west gable-end walls of model 11 during test IV, whereas both walls of model 10 developed horizontal cracks during test III (fig. 8.16). Vertical cracks developed during test IV in both retrofitted and unretrofitted models.

Damage to the in-plane walls was also affected by the stiffening and strengthening effect of the roof diaphragm. The in-plane walls of model 10 developed tension cracks at the ends of the walls due to the

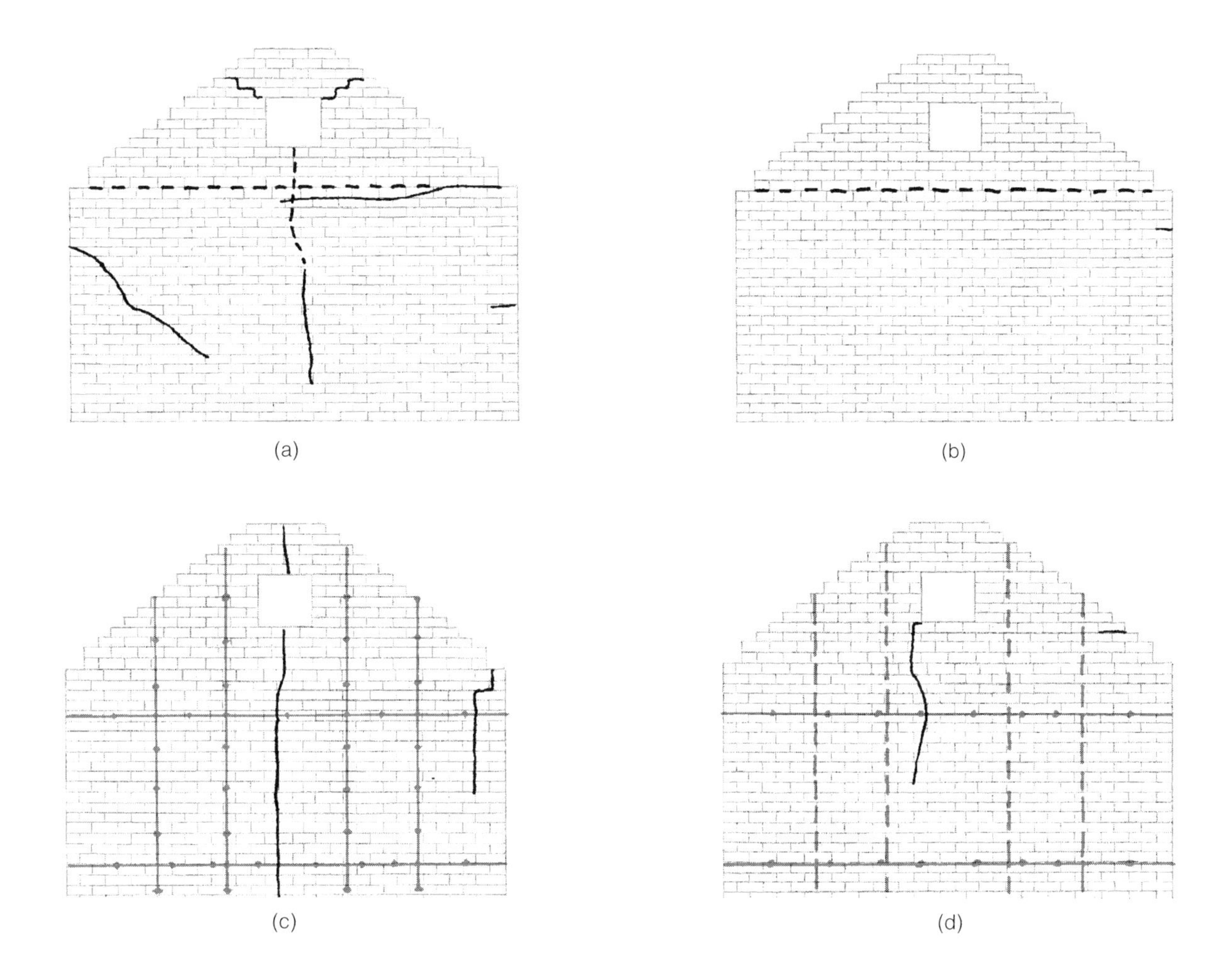

Figure 8.16
Comparison of crack patterns after test IV in out-of-plane walls for (a) model 10, unretrofitted east wall; (b) model 10, unretrofitted west wall; (c) model 11, retrofitted east wall; and (d) model 11, retrofitted west wall. Solid lines indicate cracks that occurred during test IV; dashed lines indicate cracks that typically occurred during test III.

Cracks during this test
Preexisting cracks
Straps
Center cores

pull-out forces exerted on them by the gable-end walls. The gable-end walls were beginning to pull away from the in-plane walls, which led to crack damage in the upper corners of the in-plane walls (fig. 8.17a, b). The in-plane walls of model 11 did not develop these cracks because of restraint on the dynamic motions of the out-of-plane walls provided by the diaphragms.

In the in-plane walls of the retrofitted model, the only crack that developed during test IV was the diagonal crack that extended from the upper right corner of the doorway in the north wall of model 11 (fig. 8.17c). This crack occurred because the wall already had sustained crack damage and was subjected to greater loads from the roof and floor diaphragms, which transferred loads from the out-of-plane walls to the in-plane walls. It is not possible to determine the extent to which each of these factors contributed to the development of these cracks. During test V, a small in-plane crack also developed at the upper right corner of the doorway in the south wall (fig. 8.19d). This crack turned upward and terminated when it intersected the vertical center-core bar. Nevertheless, the development of a shear crack in the south wall indicated that the loading in the in-plane walls of the retrofitted model was greater than the loading in the corresponding walls of the unretrofitted model.

Crack damage to the out-of-plane walls continued to worsen during tests V and VI as shown in figures 8.18 and 8.20, respectively. During test V, a horizontal crack developed in the east wall of model 11

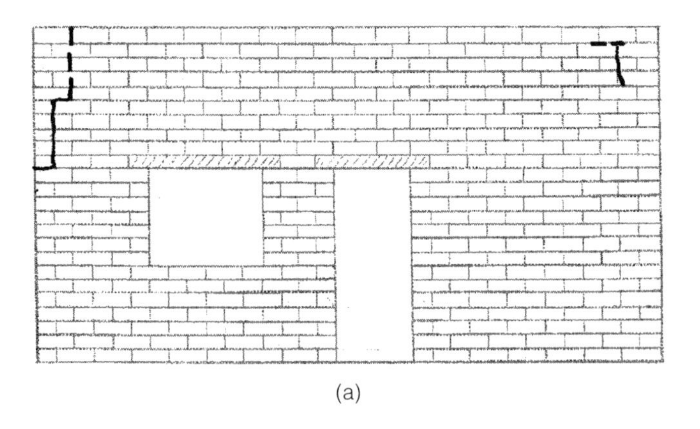

(a)

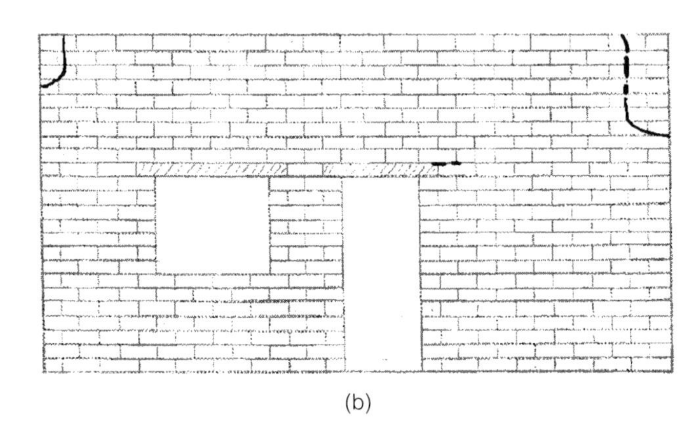

(b)

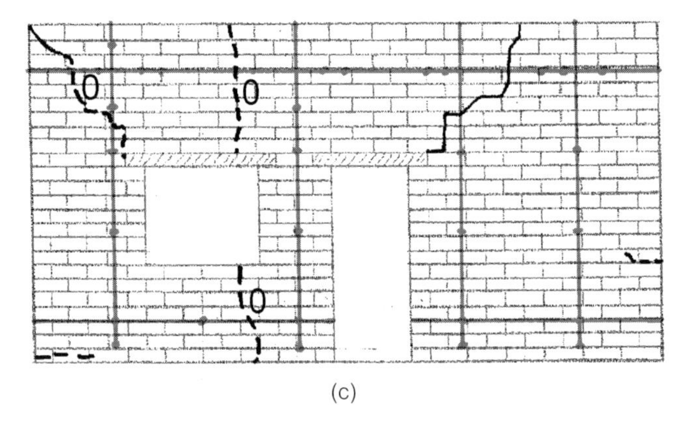

(c)

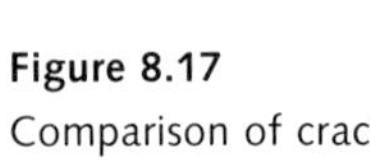

Figure 8.17

Comparison of crack patterns after test IV in the in-plane walls for (a) model 10, unretrofitted north wall; (b) model 10, unretrofitted south wall; (c) model 11, retrofitted north wall; and (d) model 11, retrofitted south wall. Solid lines indicate cracks that occurred during test IV; dashed lines indicate cracks that typically occurred during test III; cracks labeled "0" occurred before the start of the test series.

Cracks during this test

Preexisting cracks

Straps

Center cores

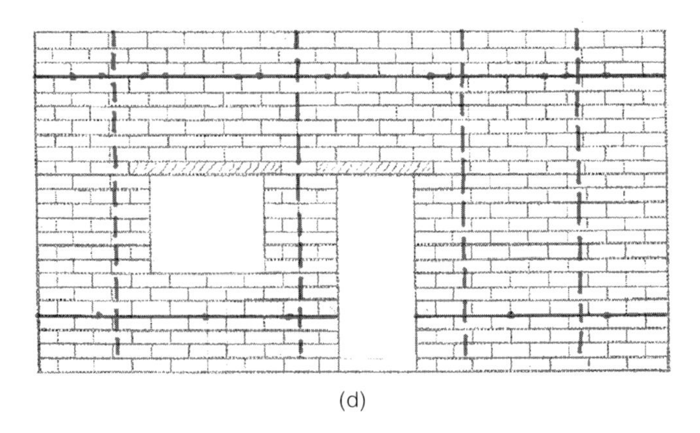

(d)

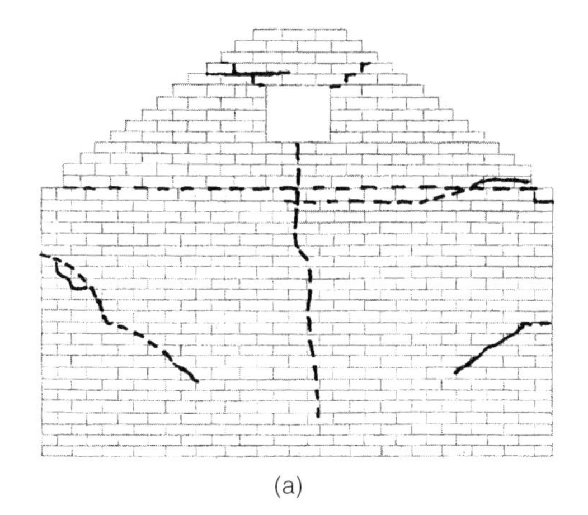

(a)

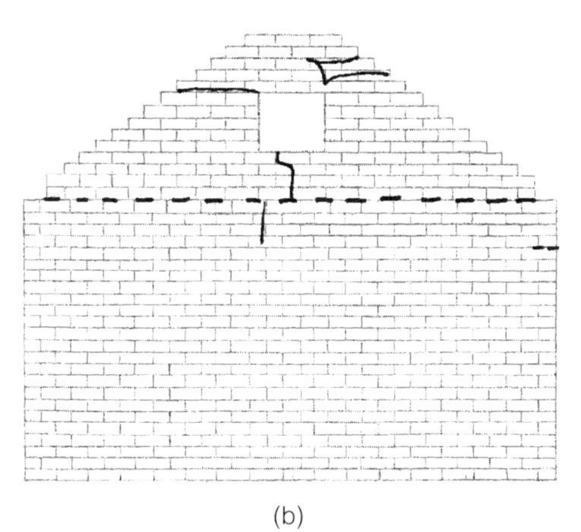

(b)

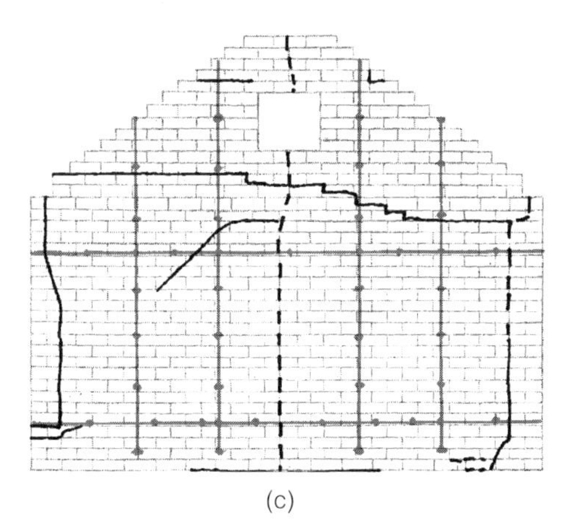

(c)

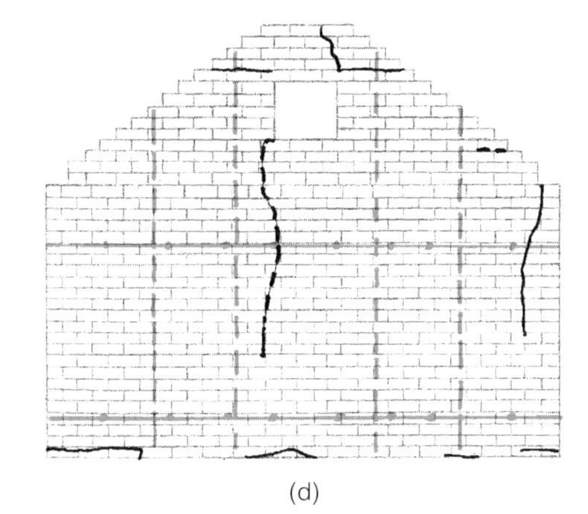

(d)

Figure 8.18
Comparison of crack patterns after test V in out-of-plane walls for (a) model 10, unretrofitted east wall; (b) model 10, unretrofitted west wall; (c) model 11, retrofitted east wall; and (d) model 11, retrofitted west wall.

Cracks during this test

Preexisting cracks

Straps

Center cores

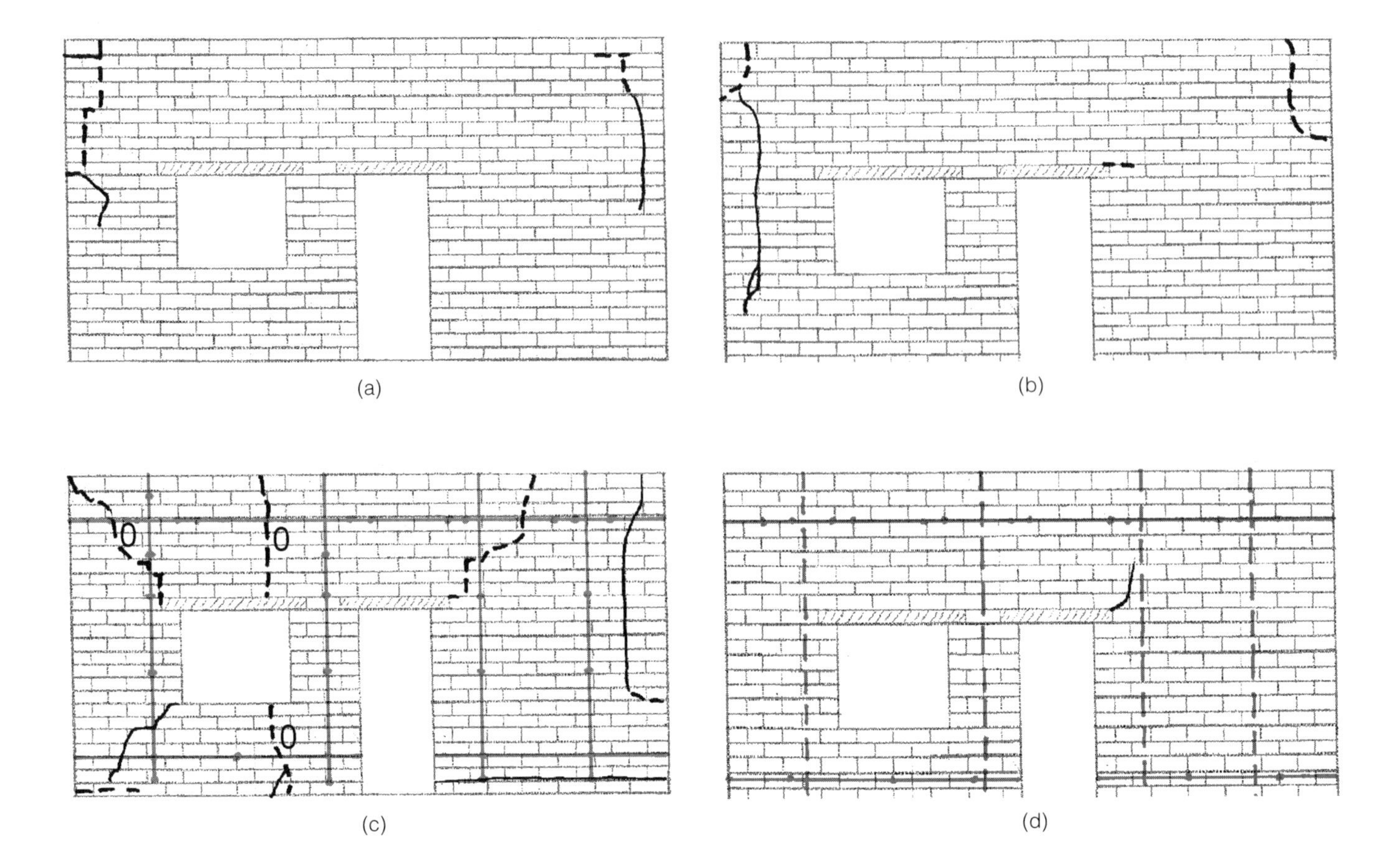

Figure 8.19
Comparison of crack patterns after test V in the in-plane walls for (a) model 10, unretrofitted north wall; (b) model 10, unretrofitted south wall; (c) model 11, retrofitted north wall; and (d) model 11, retrofitted south wall.

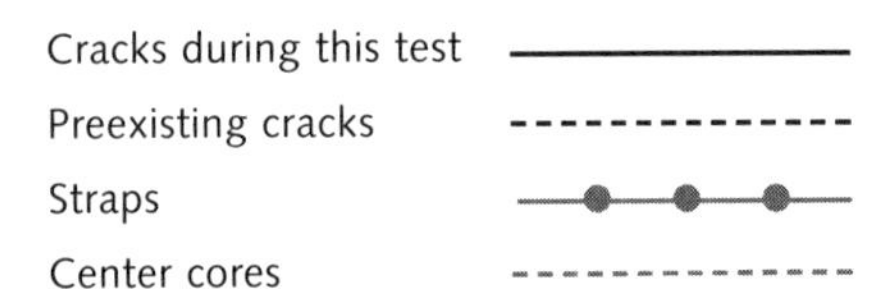

(fig. 8.18c). The roof system restrained the top of the wall, but the vertical straps were not able to prevent this type of damage. The crack damage to the east wall of models 10 and 11 was similar though not identical (fig. 8.20a, c). The vertical and horizontal strap system had no significant effect on the elastic behavior and crack development; it had an observable effect only after cracks had developed and displacements were large. Nevertheless, the effect of the strapping system is very significant. Stiffness of the roof diaphragm affected all phases of dynamic performance.

The crack damage to the west wall of model 11 shows the significant effect of the center-core rods. Through test VI (fig. 8.20), the only long cracks in the west wall were the vertical crack at the center of the wall and the vertical crack on the right (south) side of the wall. Other cracks in the wall were small and not significant. The center-core rods prevented the development of any horizontal or diagonal cracks through test VI.

Crack damage to the in-plane walls continued to progress in both walls of model 11 during tests V and VI (figs. 8.19 and 8.21). In model 11, the damage to the north wall worsened considerably, whereas the crack damage to the south wall propagated to the reinforcing bars, where the cracks terminated. In model 10, crack damage was limited to the ends of the walls and was caused by tensile stresses, not in-plane shear stresses. The differences in behavior are due largely to the additional loading that resulted from the anchorage to the roof and floor diaphragms.

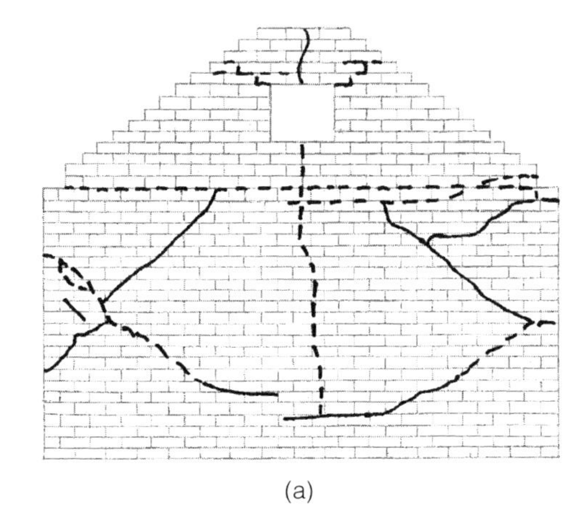

(a)

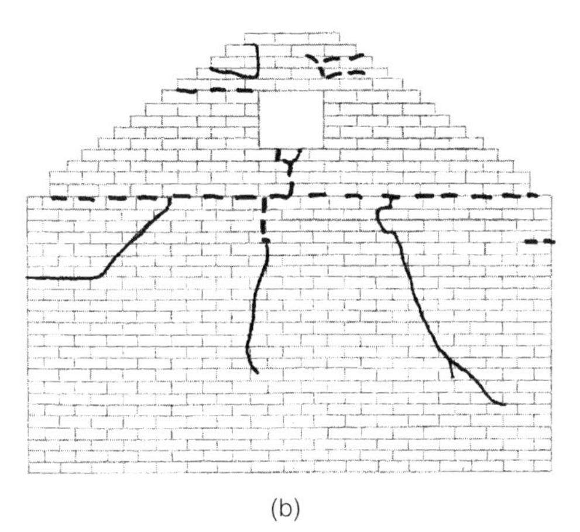

(b)

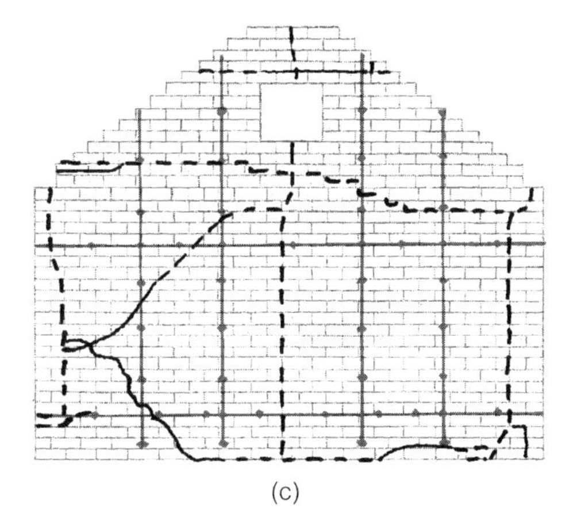

(c)

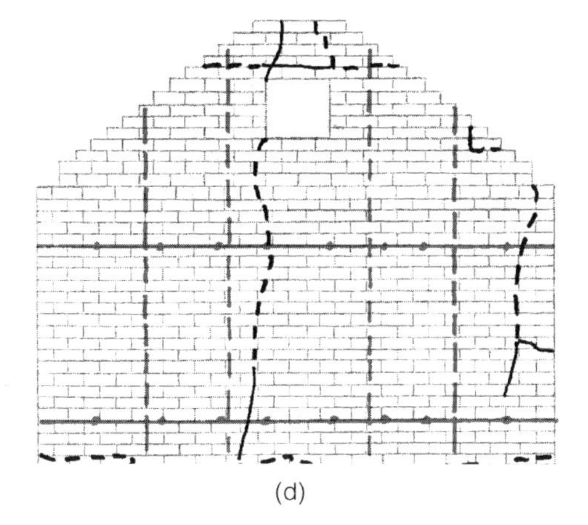

(d)

Figure 8.20
Comparison of crack patterns after test VI in out-of-plane walls for (a) model 10, unretrofitted east wall; (b) model 10, unretrofitted west wall; (c) model 11, retrofitted east wall; (d) model 11, retrofitted west wall.

Cracks during this test

Preexisting cracks

Straps

Center cores

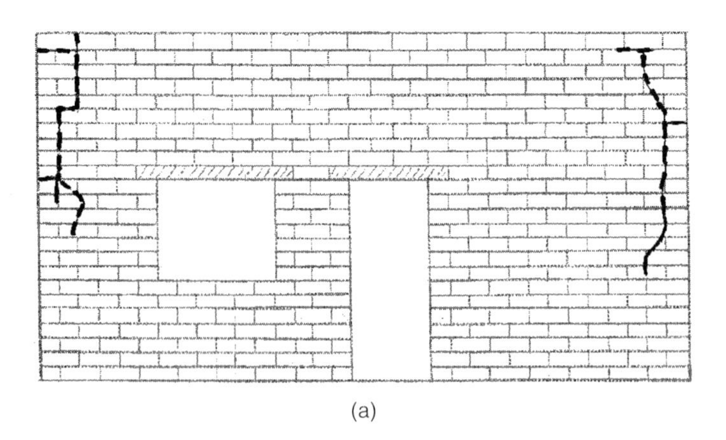

(a)

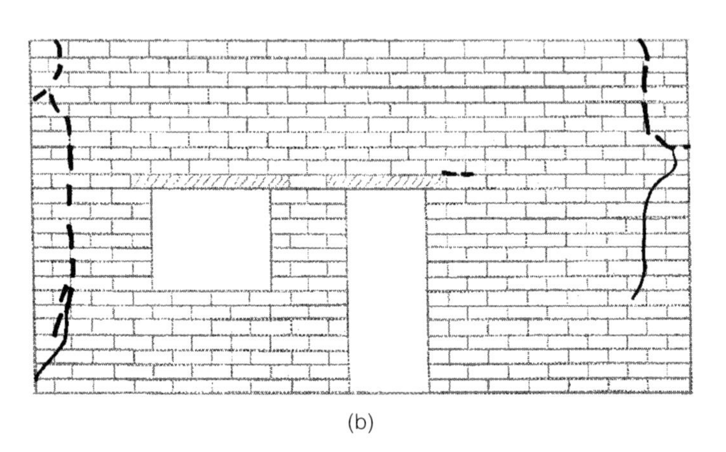

(b)

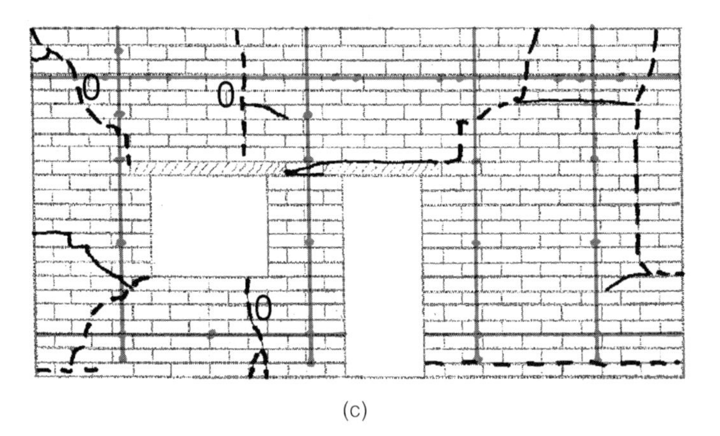

(c)

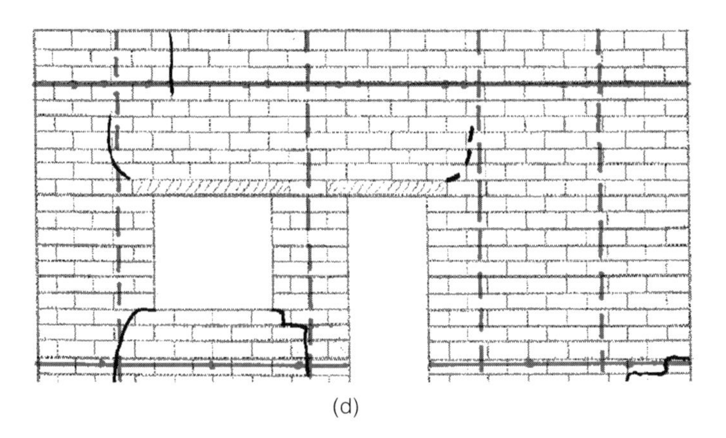

(d)

Figure 8.21

Comparison of crack patterns after test VI in the in-plane walls for (a) model 10, unretrofitted north wall; (b) model 10, unretrofitted south wall; (c) model 11, retrofitted north wall; and (d) model 11, retrofitted south wall.

Cracks during this test

Preexisting cracks

Straps

Center cores

The effect of the retrofit system can be seen in the relative out-of-plane displacements of the top of the west gable-end walls of models 10 and 11 during tests V–VII (fig. 8.22). The displacements at the middle of the west wall (fig. 8.23) were smaller than those in the east wall because the strapping system allowed greater displacements than those in the wall retrofitted with center-core rods. The time histories show that the displacements in the retrofitted west gable-end wall were approximately one-third as large as those in the unretrofitted model as early as test V, and the difference continued through test VII.

The opposite was true for the displacements of the in-plane north and south walls (figs. 8.24 and 8.25, respectively), in which the displacements of model 11 were slightly larger than those of model 10. Displacements were measured at the west end of each wall and therefore are not completely representative of the entire wall but may reflect those of cracked wall sections at the ends of the north and south walls. This cracking can be attributed largely to the additional loading transferred to these walls through the diaphragm.

Comparison of performance during severe ground motions

The out-of-plane walls of model 10 were potentially unstable during test VI and would most likely have collapsed during test VII if it were not

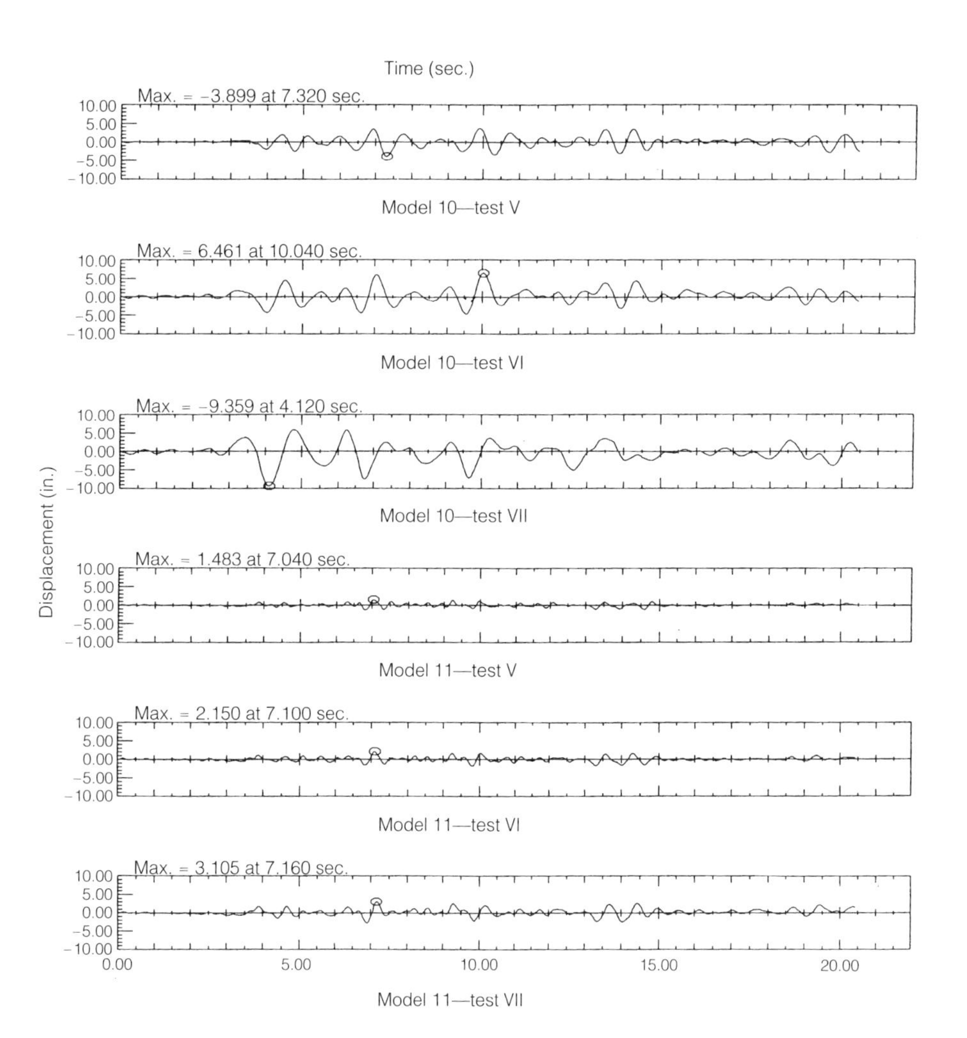

Figure 8.22
Comparison of displacement measurements at the top of the west gable-end walls for models 10 and 11 during tests V, VI, and VII.

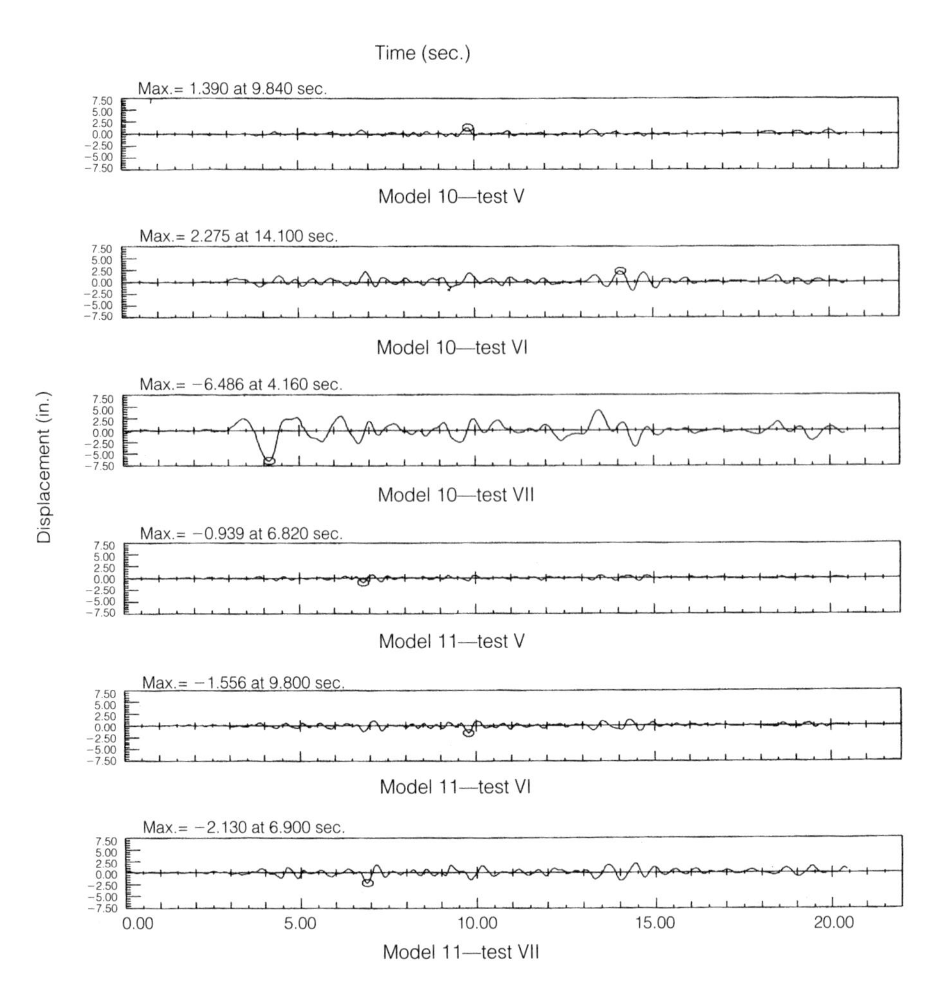

Figure 8.23
Comparison of displacement measurements at the middle of the west gable-end walls for models 10 and 11 during tests V, VI, and VII.

for the restraint at the top of the wall provided by the roof system. The dynamic motions led to large crack openings and rocking of the upper section of the wall.

In contrast, the strengthened roof system of model 11 greatly limited the displacements at the top of the wall. The maximum out-of-plane deflection at the top of the gable-end wall of model 11 was only one-third that observed in model 10 during test VII. The dynamic motions of the east wall (with vertical straps) of model 11 were larger at the center of the wall than at the top, based on visual observations of the videotape recordings. The dynamic motions of the west wall were evenly distributed over the height, with the largest displacements occurring at the top of the gable-end wall due to the stiffening effects provided by the center-core rods. The displacements at the middle of the gable-end wall (at about two-thirds height location) were approximately two-thirds as large as the displacement at the top of the wall.

The out-of-plane displacements at the top of the gable-end walls for model 11 were significantly smaller than those for model 10. The displacement of the west wall of model 10 during test VII was nearly 23.9 cm (9.4 in.). Because the walls collapsed near the start of the following test, a slightly larger displacement was required to cause instability. Therefore, the displacement for instability can be assumed to be

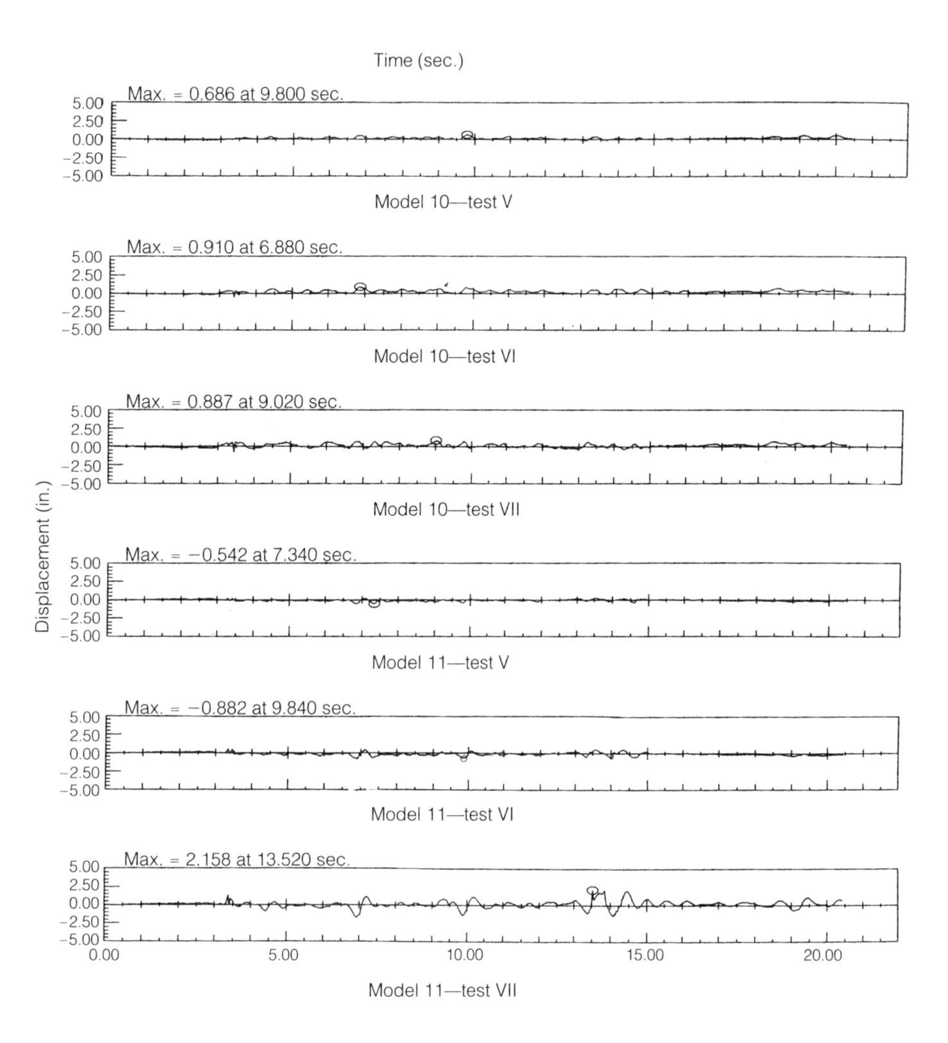

Figure 8.24
Comparison of displacement measurements of the north walls for models 10 and 11 during tests V, VI, and VII.

approximately 25.4 cm (10 in.). For the retrofitted model, the largest displacement at the top of the gable-end wall during test VIII* was nearly 8.9 cm (3.5 in.). This displacement value is only about one-third of the value required for instability to occur.

The response of these model buildings can be used to estimate the safety factor provided by the retrofit system with regard to the risk of overturning of the out-of-plane walls. Since the dynamic level of displacement required to cause instability of the 40.6 cm (16 in.) thick walls is at least 25.4 cm (10 in.), the safety factor for overturning was about a factor of 3.

Photographs of the damage to each wall after test VII appear in figures 8.26–8.29. Crack patterns are shown in figures 8.30 and 8.31. Damage to the out-of-plane walls was significant after test VII in both walls of model 10 and in the east wall of model 11 (fig. 8.30a–c). Although the crack pattern was somewhat different, the extent of the cracks and the amount of spalling was similar in both models. The vertical straps had little effect on the extent of damage; their principal purpose was to reduce damage progression and prevent instability. The extent of damage to the west wall of model 11 was the least of all four walls because of the ductility and strengthening provided by the center-core rods (fig. 8.27).

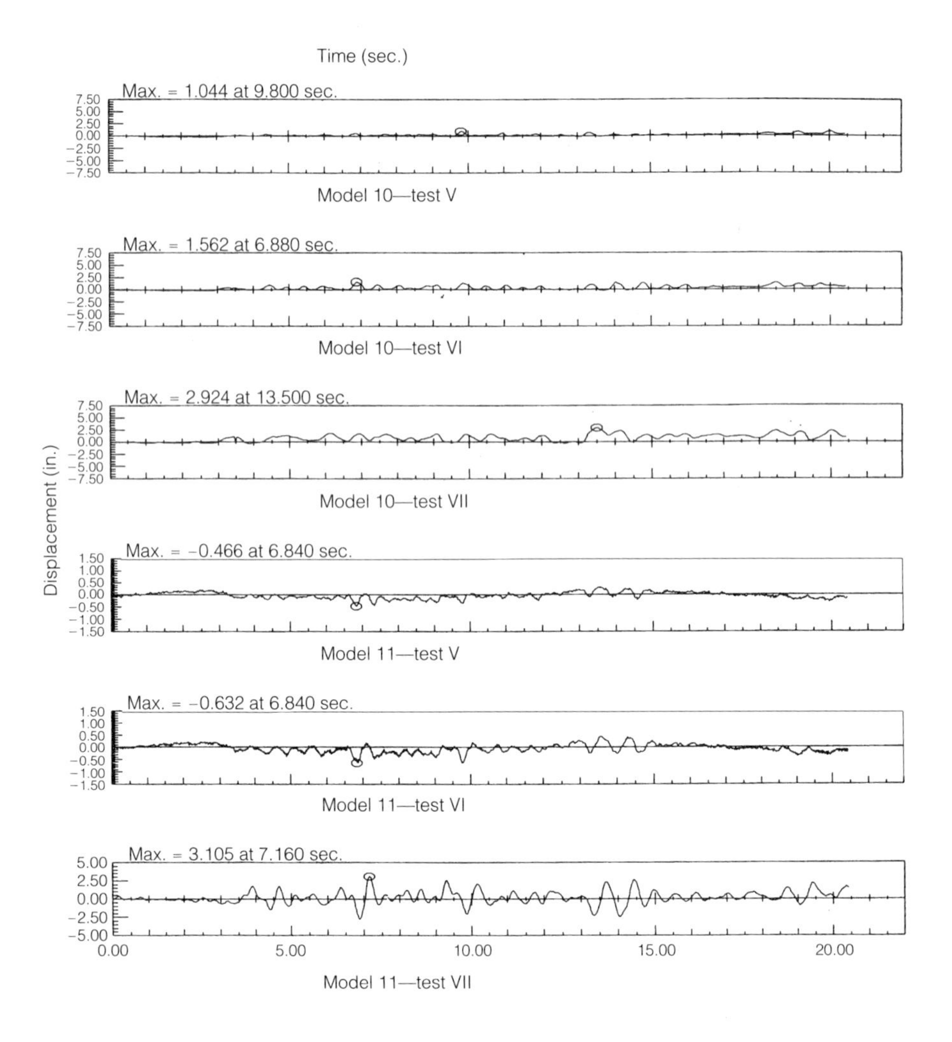

Figure 8.25
Comparison of displacement measurements of the south walls for models 10 and 11 during tests V, VI, and VII.

During the severest tests, additional cracks developed in the in-plane north and south walls (figs. 8.28, 8.29, and 8.31). For the most part, the crack damage in these walls was relatively minor. One exception was the section to the east side of the window at the end of the north wall of model 11, where diagonal cracks developed at the top and bottom

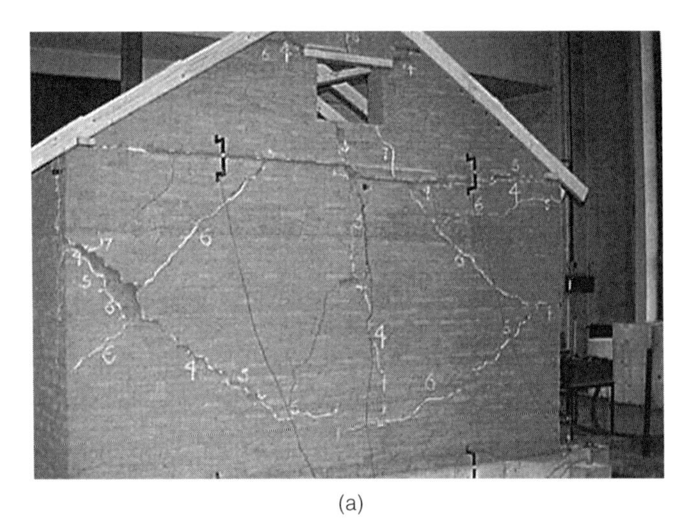

(a)

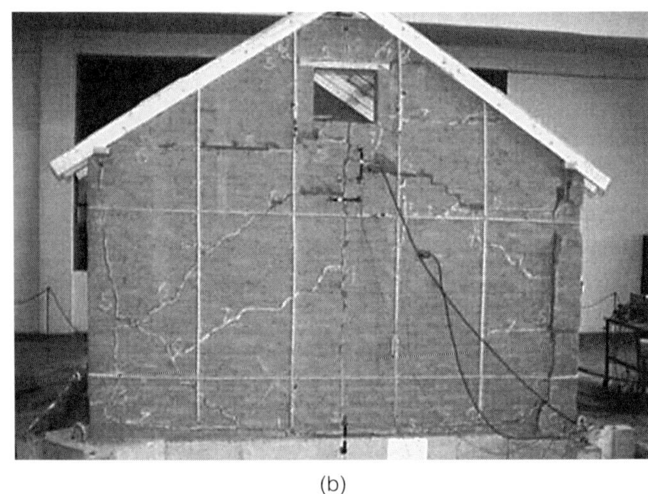

(b)

Figure 8.26
Comparison of damage to the east walls for (a) model 10 and (b) model 11 after test VII.

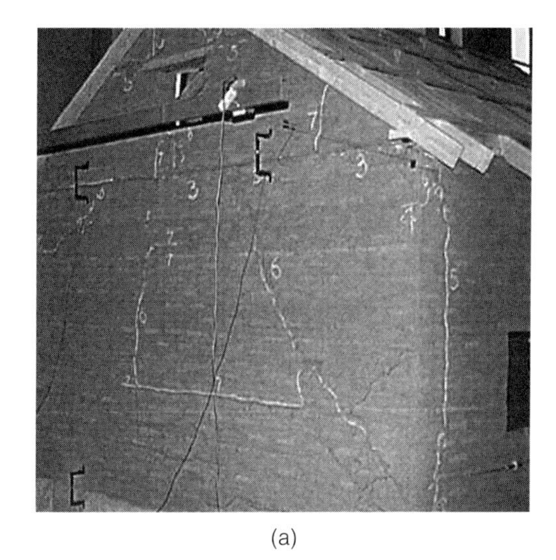

(a)

(b)

Figure 8.27
Comparison of damage to the west walls for (a) model 10 and (b) model 11 after test VII.

corners of the window. The section was bounded by two large cracks and was not restrained by an adjacent wall. The lower diagonal crack provided a slip plane for this block. Movements along the slip plane were aided by the dynamic motions of the shaking table and gravity. This type of damage has been observed both in the laboratory and in the field and represents a common condition that needs to be addressed by a complete retrofit system. The center-core rod near the southeast corner (fig. 8.29d) was effective in anchoring this section. Shear cracks began to develop in this corner that were similar to the crack at the northwest corner; the center-core rods prevented the full propagation of the upper and lower cracks. Even if the cracks had propagated more fully, the center-core rod would have acted as a dowel pin across the joint and greatly limited the amount of offset that could occur at this location.

A second exception to the minor damage of in-plane cracks occurred in the south wall of model 10, as illustrated in figure 8.29a. The crack shown on the left in this photograph is a tension failure caused by the out-of-plane motions of the west wall and is not a typical shear crack.

(a)

(b)

Figure 8.28
Comparison of damage to the north walls for (a) model 10 and (b) model 11 after test VII.

Figure 8.29
Comparison of damage to the south walls for models 10 and 11 after test VII, showing (a) model 10, south wall, west side; (b) model 10, south wall, east side; (c) model 11, south wall, west side; and (d) model 11, south wall, east side.

Performance Observations and Summaries

Crack initiation in the out-of-plane walls of both model buildings began during tests III and IV, when peak table accelerations were 0.11 g and 0.17 g, respectively. The only cracks in the in-plane walls, besides tension cracks at the ends of the walls, did not occur until test VII in model 10, when the peak table acceleration was 0.47 g. Given the flexible roof system and the absence of interconnections between the walls, the loading on the in-plane walls was due primarily to the dynamic forces caused by the acceleration of the mass of the walls themselves. Based on simple force-resistance calculations, the in-plane shear walls failed when the shear stresses were at least 3 psi (2.06 N cm^{-2}). Additional loads may have been transferred from the out-of-plane walls, and therefore this stress level is a minimum value. This analysis is based on the observation that the in-plane shear cracks in model 10 did not occur until test VII, when the PGA was 0.47 g. The calculation of mass is minimized by considering only the mass of the walls above the door openings; the area used is the gross minimum area at the window opening. No compensation is made for the relative stiffness of the wall panels, and the self-weight of the shear walls is the only mass used in this analysis.

In the out-of-plane walls, full crack development occurred during tests V (PGA = 0.23 g) and VI (PGA = 0.27 g). (Fig. 8.30 shows crack patterns in these walls after test VII.) Full crack development did not occur in the in-plane walls, except perhaps in the north wall of model 11 after test VIII*. (Fig. 8.31 shows crack patterns in these walls after test VII.) Full crack development is the point at which all the edges of the major sections of a wall are defined by cracks. A drawing of the crack pattern in the out-of-plane east wall of model 11 after test VI is shown in figure 8.32a. The major cracked sections or "blocks" of this wall are labeled B1 through B5. Each section is bounded by a free surface or by a crack on all four sides. Despite the increased intensity of the table motion in test VII, the only major crack that developed was the one that split B4 into two sections (fig. 8.32b). Once cracks developed,

Figure 8.30

Comparison of crack patterns after test VII in the out-of-plane walls for (a) model 10, unretrofitted east wall; (b) model 10, unretrofitted west wall; (c) model 11, retrofitted east wall; and (d) model 11, retrofitted west wall.

Cracks during this test
Preexisting cracks
Straps
Center cores

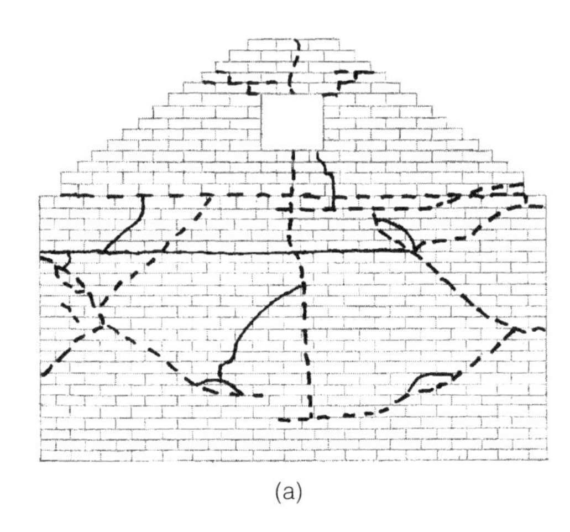

(a)

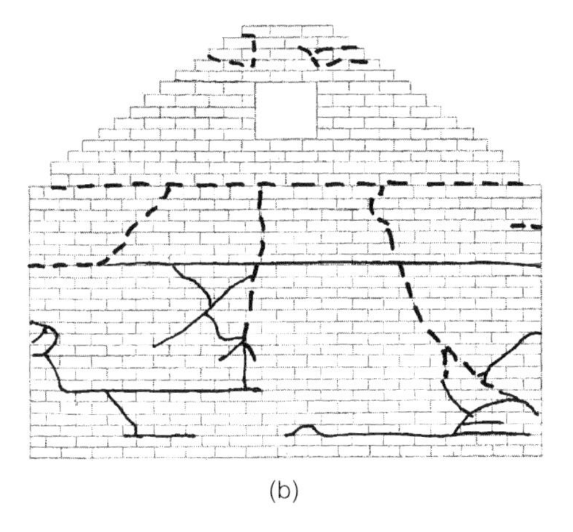

(b)

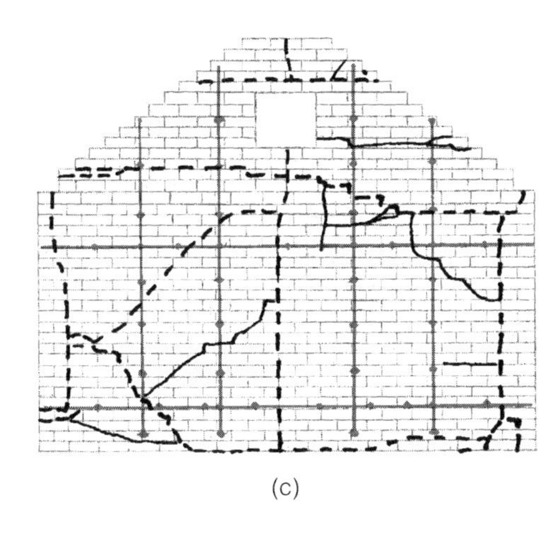

(c)

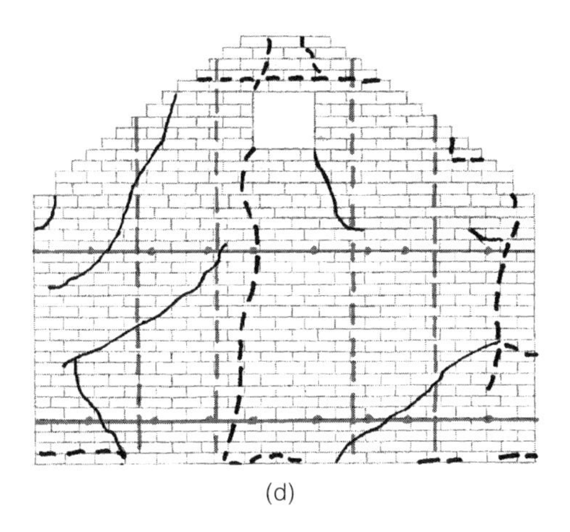

(d)

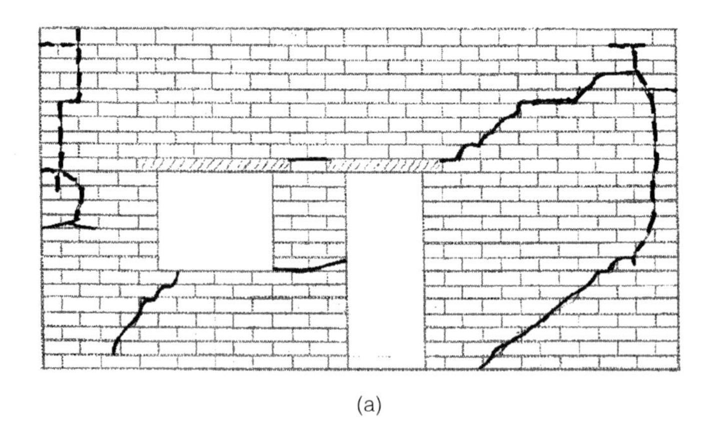

(a)

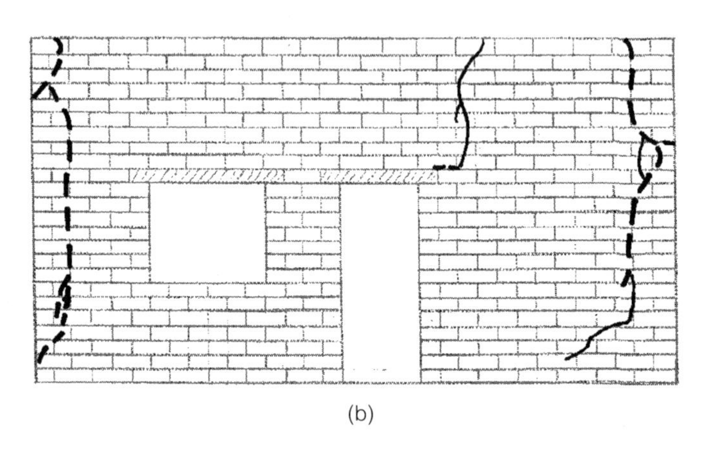

(b)

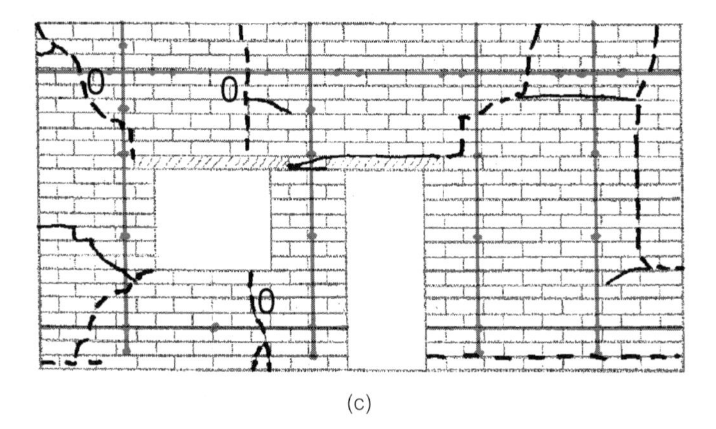

(c)

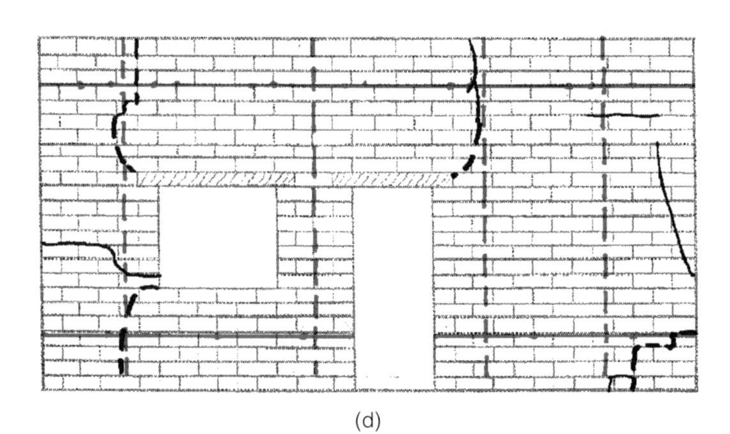

(d)

Figure 8.31

Comparison of crack patterns after test VII in the in-plane walls for (a) model 10, unretrofitted north wall; (b) model 10, unretrofitted south wall; (c) model 11, retrofitted north wall; and (d) model 11, retrofitted south wall.

Cracks during this test

Preexisting cracks

Straps

Center cores

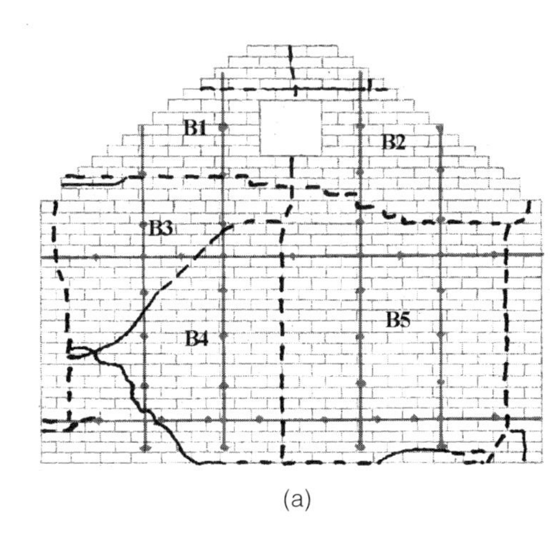

(a)

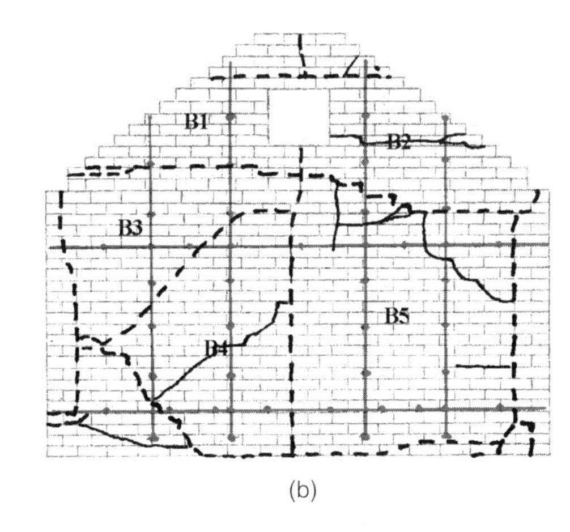

(b)

Figure 8.32
Crack pattern development in east wall of model 11 (a) after test VI and (b) after test VII. During test VI, the wall cracked into five major sections, or blocks, labeled B1 through B5. During test VII, three of the major blocks remained intact, but B2 and B4 each cracked into two sections.

Cracks during this test
Preexisting cracks
Straps
Center cores

the primary dynamic behavior of the building was determined by the dynamic motions of the cracked wall sections. In this case, it is difficult to apply enough force to a block to cause it to crack further. Additional damage was largely caused by gradual degradation of the contact areas between each of the major cracked wall sections.

The strapping system used on the east wall of model 11 was effective in providing wall stability. A drawing of the crack pattern after test VII and photographs showing the condition of the walls are presented in figure 8.33. Cracks and spalling were significant, and an offset of almost 5.1 cm (2 in.) occurred at the northeast corner (fig. 8.33c).

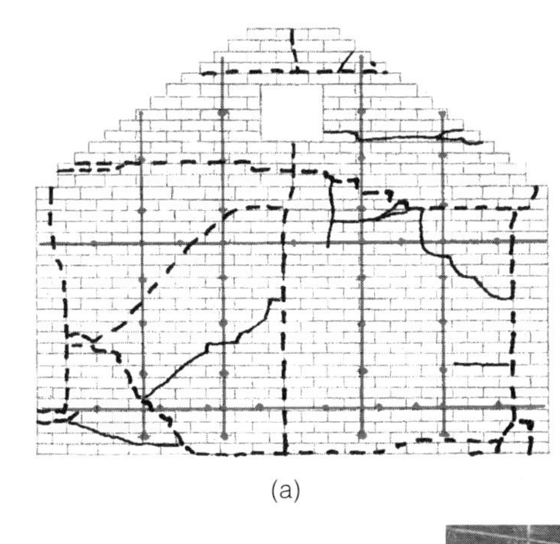
(a)

(b)

(c)

Figure 8.33
Damage to model 11, east wall (retrofitted with vertical straps) after test VII, shown in (a) diagram of crack pattern, with vertical lines indicating position of straps; (b) photograph of strapped wall; and (c) detail of north side of wall.

Cracks during this test
Preexisting cracks
Straps
Center cores

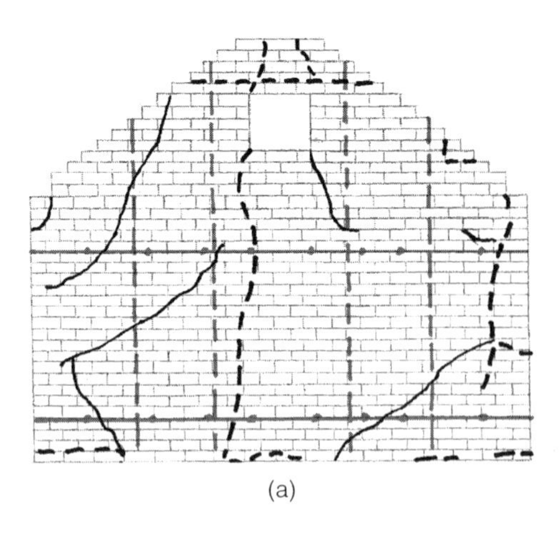

(a)

(b)

(c)

Figure 8.34
Damage to model 11, west wall (retrofitted with center-core rods) after test VII, shown in (a) diagram of crack pattern; (b) photograph of north side of wall; and (c) photograph of south side of wall.

Cracks during this test
Preexisting cracks
Straps
Center cores

Although the east wall of model 10 collapsed, with failure occurring at about the two-thirds height level, the east wall of model 11 was stable even after the large displacement during test VIII*.

The center-core system used on the west wall of model 11 performed extremely well. The condition of the wall after test VII is shown in figure 8.34. Although the full crack patterns developed during tests VI, VII, and VIII*, the wall was stabilized by the center-core rods. There were no large cracks, offsets, or spalling on this wall; despite the cracks, the wall remained in very good condition.

The strapping system of model 11 was not effective in controlling the in-plane diagonal cracking at the east end of the north wall (fig. 8.35). The upper and lower straps were highly stressed, which resulted in some damage to the adjacent adobe. Nevertheless, the straps continued to perform well throughout the testing program. Some additional retrofit measures need to be installed to minimize this type of damage, which could lead to the collapse of the corner section of the wall.

The center-core elements in the south wall of model 11 were extremely effective in controlling the initiation and progression of crack damage (fig. 8.36). Three of the cracks that started at the corners of the openings in the wall were arrested when they reached the center-core rods. In the later tests, the cracks continued to progress but never became very large.

The roof system installed on the models had adequate stiffness to prevent large top-of-the-wall displacements, and the anchorage at the tops of the walls minimized damage in these areas. The ties from the floor diaphragm to the horizontal strap at the floor level performed well. No failures of any of these connections were observed. The through-ties that connected the straps on either side of the walls also performed well throughout.

Summary of Comparison of Performance of Small- and Large-Scale Models

Because the small-scale (1:5) models were not instrumented to record displacements and accelerations, direct quantitative comparison with

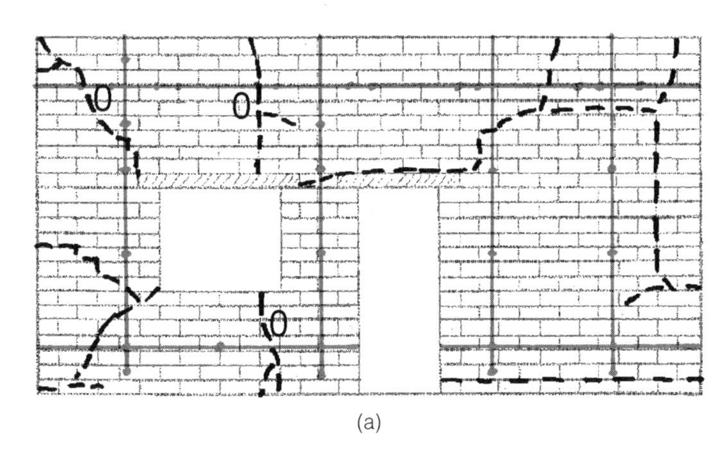

(a)

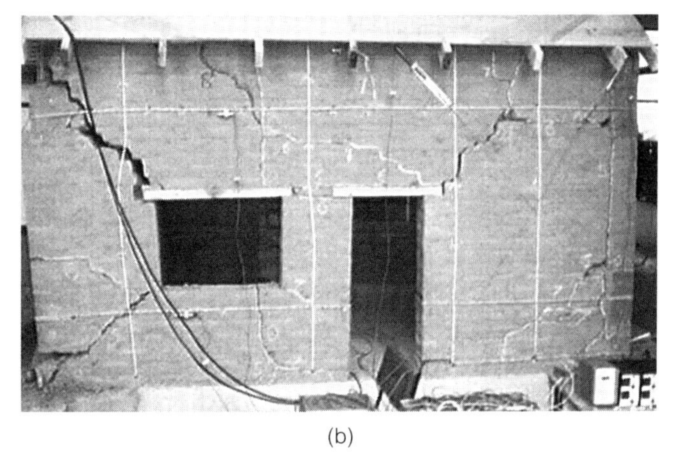

(b)

(c)

Figure 8.35
Damage to model 11, north wall (retrofitted with vertical straps), shown in (a) diagram of crack pattern after test VII; (b) photograph after test VIII*; and (c) detail of east side of wall after test VIII*.

Cracks during this test
Preexisting cracks
Straps
Center cores

large-scale (1:2) model results was not possible. However, comparison of crack initiation and development, offsets at cracks, failure modes, and other aspects of the relative dynamic performance of the two sets of structures can provide the basis for an evaluation of the effectiveness of the small-scale tests.

(a)

(b)

Figure 8.36
Damage to model 11, south wall (retrofitted with center-core rods) after test VIII*, showing (a) west side and (b) east side of wall.

The elastic performance and early crack development modes of the small- and large-scale models were similar. The random development of cracks, the minor differences in shaking-table motions, and the variation in adobe materials properties could have accounted for the minor behavioral differences among the tapanco-style models.

The types of behavior most affected by the scale of the models became apparent at the higher displacement and acceleration levels. The most significant of these were out-of-plane overturning in the small-scale models and offsets resulting from in-plane diagonal cracks in the large-scale models. The risk of overturning failure may have been overstated by the performance of the small-scale models, however. Because the mass of the wall is underrepresented at the 1:5 scale, and because greater masses resist overturning, this effect may have occurred at a lower ground motion level than was found to occur in the 1:2 scale models. In full-scale buildings, the high vertical gravity loads would tend to inhibit overturning.

Problems associated with damage to in-plane walls along diagonal cracks become more severe as the scale increases. Because of the greater mass of the wall and the fact that the displacements along diagonal cracks are cumulative, the effects produced may be more significant than those resulting from restrictions to movement caused by increased friction between the moving blocks. Therefore, larger-scale models will tend to exhibit greater offset damage than smaller-scale models, and this failure mode is expected to be serious in unretrofitted, full-scale buildings. Despite these differences, the overall performances of the small- and large-scale models were very similar, and the retrofit measures shown to be effective on the small-scale models were also effective on their large-scale counterparts.

Chapter 9

Recent Application of GSAP Technology

A recent systematic application of GSAP seismic retrofit technology was completed in the summer of 1998 at the Del Valle Adobe at Rancho Camulos, located near Piru, California. This building is a rancho of the Mission San Fernando and is considered an outstanding example of the style of old California ranchos. The main residence served as a model for the home of Ramona, the title character of Helen Hunt Jackson's popular nineteenth-century romance novel, noted for its portrayal of the idyllic, pastoral days of early California. The adobe building was damaged extensively during the 1994 Northridge earthquake (EPGA 0.3–0.4 g) (Tolles et al. 1996). Two walls of Ramona's bedroom collapsed, and the adjacent gable-end wall was severely damaged but did not collapse (fig. 9.1a, b). Crack damage occurred throughout the building, especially at corners and—because of the pounding of perpendicular walls against each other—at wall intersections also. Delamination of interior and exterior plaster was extensive, as was spallation of adobe in areas weakened by previous repeated exposure to water.

(a)

(b)

Figure 9.1
Ramona's room, Del Valle Adobe at Rancho Camulos, Piru, California, after the 1994 Northridge earthquake, showing (a) southeast corner of building and (b) west wall of room. Buttresses on the south wall prevented the total collapse of that wall.

In many locations, the walls had pulled away from the ceiling joists, and damage to the walls further reduced their ability to support the joists. Three major physical conditions—water damage, preexisting cracks, and material weakness—influenced the performance of the Del Valle Adobe during the Northridge earthquake. The most significant was moisture damage. There had been water damage to the lower adobe walls and foundation in several locations. Repeated wet/dry cycling has resulted in deteriorated and weakened adobe at the base of many walls. To hide the moisture damage, the lower section of the adobe walls had been covered in many areas with large concrete patches, which forced the water to rise even higher in the walls.

The adobe material used at Rancho Camulos is a silty, sandy soil with a relatively small amount of clay. As a result, the walls cracked at relatively low levels of earthquake excitation. The building also had many cracks throughout the structure that occurred prior to the earthquake. Judging from the nature of these cracks, it is likely that many occurred during previous earthquakes. The larger cracks had been filled with concrete and covered with plaster. No seismic retrofitting had been done prior to the earthquake. Although there are three stone buttresses outside the corner bedroom (Ramona's room) and one on the other west elevation of the west wing, it is likely that these were constructed to stabilize the walls for other reasons, such as moisture damage to the lower wall. There is also a pair of wooden pilasters anchoring a tie rod that was probably installed in response to wall leaning. For the most part, there were no structural elements that could tie the walls together or tie the roof-ceiling system to the walls.

At the request of rancho owners, a damage assessment was made and a retrofit and repair strategy was proposed by a private design team headed by one of the authors (Tolles). The team worked with the California State Office of Historic Preservation and the Ventura County Department of Buildings and Safety to ensure that the retrofit design conformed to the U.S. Secretary of the Interior's Standards for Rehabilitation of Historic Buildings and requirements of existing local building codes.

The design of the retrofits for Rancho Camulos was based largely on the results of GSAP research.[1] A careful review of the proposed retrofit measures was carried out by the designers and state and county officials. The measures that have been approved and installed included placing horizontal stainless-steel cables or nylon straps around perimeter walls, which in some cases were anchored to ceiling joists (fig. 9.2a); installing vertical pre-tensioned steel cables on both sides of thin walls ($S_L > 8$), or on walls that were particularly vulnerable because of previous earthquake damage (fig. 9.2b, c); inserting epoxy-grouted vertical center-core rods in newly constructed adobe walls; and installing appropriate anchorage between the walls and roof at the floor and ceiling levels.

[1]Other California adobes whose retrofits were influenced in the past by some of the GSAP concepts were Courthouse Adobe, Shafter; O'Hara Adobe, Toluca Lake; Lydecker Adobe, Aptos; Jameson Adobe, Corona; and the Mission San Gabriel convento, San Gabriel.

(a)

(b)

(c)

Figure 9.2
Installation of steel cables at the Del Valle Adobe, showing (a) horizontal cable anchored to ceiling joists; (b) vertical cable installed in previously damaged walls; and (c) enlarged detail of cable anchor (area of enlargement shown in b).

This first implementation of all the principal retrofit techniques studied in GSAP required some redesign of laboratory-tested details for application to real-world conditions, as well as review and acceptance by architectural and engineering building officials and officials of the California State Office of Historic Preservation. The project also required input and review by the buildings' owners, who were particularly concerned with preserving the historic features and maintaining safety in and around the buildings.

Chapter 10
Summary and Conclusions

The goal of the Getty Seismic Adobe Project, as stated in the project's first annual report (Thiel et al. 1991), was "to develop technical procedures for improving the seismic performance of existing monumental adobe structures consistent with maintaining architectural, historic, and cultural conservation values."

Although the application to monumental adobe structures has only begun, the tests performed and the results obtained demonstrate that retrofit measures can be used effectively to provide life safety by minimizing the likelihood of catastrophic collapse while having only a minimal impact on the historic fabric of the retrofitted building. In the years prior to the initiation of GSAP, some gains had been made by the engineering community in developing retrofit measures that were sensitive to the issue of preservation of historic building fabric. Nevertheless, there was very little information about how these less intrusive retrofit systems would, in fact, behave.

The results of the GSAP research effort, involving small- and large-scale seismic simulation tests on adobe buildings, has provided the necessary technical information to justify the use of less intrusive retrofit measures.

The principal technical accomplishments of GSAP were

- Development of the basic theoretical framework for understanding seismic retrofit systems for adobe buildings.
- Development of retrofit systems that are likely to be very effective in providing life safety and in minimizing the extent of damage.
- Testing of a set of retrofit tools that can be used on historic adobe buildings.
- Presentation of reports on the project and publication of technical papers on the studies.
- Development of planning and engineering guides to assist owners, building department officials, engineers, architects, and conservators in defining the requirements and in designing retrofit systems for historic adobe buildings (Tolles, Kimbro, and Ginell n.d.).

- Documentation of the study of the relative seismic performance of historic adobe buildings affected by the Northridge earthquake on January 17, 1994 (Tolles et al. 1996). The results of this survey are significant in that they represent the behavior of real buildings during a large-scale earthquake and provide important information that can be useful in understanding the performance of historic adobe buildings and in developing retrofit techniques.

Protection of historic fabric has been a serious consideration throughout the GSAP research program. A variety of techniques were suggested and tested that might apply to the various building conditions encountered in the field. This multidisciplinary approach, coupled with the counsel provided by the GSAP Advisory Committee, makes this research program unique in its application to historic buildings.

Assessment of Retrofit Measures

The retrofit systems tested in GSAP involved horizontal and vertical straps, ties, vertical center-core rods, and improvements in the anchoring of the roof to the walls. Each method proved to be successful in reducing the tendency of the model buildings to collapse.

The retrofit method using vertical straps was most effective for reducing the risk of out-of-plane wall collapse. Vertical straps had little or no effect on the initiation and early development of crack damage. When displacements or offsets became significant, however, the strapping system controlled the relative displacement of cracked sections of walls, which, if left uncontrolled, led to instability. When coupled with tied anchorage to the roof and/or floor system, the out-of-plane overturning or mid-height collapse of walls can be prevented.

In-plane damage was much less affected by vertical straps. This is largely because in-plane offsets are smaller in magnitude and more likely to persist after the dynamic motions are completed. Straps can prevent large displacements but not small crack offsets. Straps are also useful in preventing piers from becoming unstable.

Vertical center-core rods installed in the adobe walls were found to be particularly effective in delaying and limiting the damage to both in-plane and out-of-plane walls. The initiation of cracks was delayed because of the center-core rods. Some cracks in the in-plane walls that started at the corners of the door and window openings propagated to a center-core rod and then were arrested for one or two more tests. The cracks never became severe.

In the out-of-plane walls, the center-core rods acted as reinforcing elements. Epoxy grout surrounding the rods provided effective shear transfer between the adobe and the steel rods. The epoxy soaked into the adobe unevenly because of the variation in the walls, especially at the mortar joints, and therefore provided a positive attachment between the steel rods and the adobe walls. The walls with center-core rods were then able to act as structural reinforcing elements in the vertical direction. In addition, the rods acted as dowel pins that minimized the relative motion of adobe blocks. Removal of such rods would be difficult.

One advantage of the strapping system is that the installation has minimal impact on the historic fabric of the wall and is also reversible. If surface renderings such as murals are important, then the interventions required for installation of a strapping system would not be appropriate. For most historic buildings, however, surface renderings of the walls are not particularly significant, especially in cases where there is deterioration of the plaster/stucco or damage has occurred from an earthquake. When disturbance of the wall surface is unacceptable, the use of center cores will avoid damage to wall surfaces.

Relative Model Performance

The performance of small- and large-scale model buildings was very similar in many ways. The general development of cracks, the types of cracks, and the failure modes were similar. The effects of the retrofit measures on building behavior were also similar. For the most part, the behavior of the small-scale models was an acceptable predictor of large-scale model performance.

The principal physical difference between the small- and large-scale models was in gravity loads. As a result, specific differences were seen in both the out-of-plane and in-plane wall performances. Global performances were very similar. The most outstanding difference was the occurrence of diagonal cracks in the in-plane walls. Diagonal cracks allow displacements that are cumulative and slippage that is exacerbated by vertical loads. As a result, diagonal cracks were more of a problem in the larger-scale model and would be expected to be at least as serious a problem in full-scale buildings, in which diagonal cracks near the ends of piers and walls are of particular concern.

On the one hand, overturning of walls is less of a problem in full-scale buildings because the vertical loads resist overturning. On the other hand, the condition of the base of the walls in the model buildings was very good. In many actual adobe buildings, the adobe at the base of the walls has often been weakened by exposure to moisture, and walls that are weakened at the base are highly vulnerable to overturning.

The performance of adobe buildings was demonstrated in real-world terms by the 1994 Northridge earthquake, in which many historic adobe buildings were seriously damaged. Given that this research program was under way and that there was a dearth of information regarding the past performance of earthquake-damaged adobe buildings, this destructive event did, however, provide a fortuitous opportunity for comparison of field observations, scientific analysis of actual buildings, and the results of laboratory testing.

Despite the information gained from the Northridge earthquake, there is still much that may be derived from a study of the performance of well-retrofitted adobe buildings during a major earthquake. Some of the buildings damaged during the Northridge earthquake had been partially retrofitted, but these measures were implemented in a piecemeal manner, and none provided an integrated retrofit system. More complete information on the performance of systematically retrofitted adobe buildings will be obtained after major earthquakes in the future. The development of engineering standards is based on engineering analysis, research, and field observations. Information presently available on the basic performance of adobe buildings is significant and can be used as the basis for developing standards for the performance and seismic retrofit of these important historic structures.

Chapter 11
Suggestions for Future Research

The Getty Seismic Adobe Project has not answered all of the relevant questions regarding the design of seismic retrofit systems for historic adobe buildings. Other areas for future research are

- **Connections.** This area has not been addressed in detail in the GSAP research effort, particularly for the tops of walls, where durable connections are difficult to make. Large loads at connections into adobe walls can result in additional damage. *Planning and Engineering Guidelines for the Seismic Stabilization of Historic Adobe Structures* (Tolles, Kimbro, and Ginell n.d.) will discuss methods for predicting and planning for this damage. Nevertheless, the required strength of these connections has not been directly addressed.
- **Strong-motion instrumentation.** Data collected on the behavior of historic adobe buildings during an actual earthquake event should be useful in analyzing the performance of these structures. The interaction between existing roof systems and the adobe walls is one area where additional data are needed. Relatively few historic adobe buildings remain, and the chances of getting information any time in the near future may be low. Nevertheless, the data obtained from such an investigation may be extremely valuable and, perhaps more important, may be reassuring to those who have questions about other conclusions drawn from past research.
- **Analytical methodologies.** The work conducted through GSAP concerns the development of basic strategies for designing seismic retrofits of historic adobe buildings. More extensive work is necessary to turn these methods into analytical procedures. The analysis should include the complex interaction between a thick adobe wall rocking out of plane and supported by a flexible diaphragm (either plywood, straight sheathing, or a partial diaphragm) and

in-plane walls that have suffered crack damage. This complex system could be addressed either analytically or by using information based on actual test results.

The GSAP research effort has provided significant additions to the body of knowledge on the behavior of adobe structures retrofitted according to stability-based design principles. Much work remains to be done in the future; for now, at least, a foundation exists for the design of seismic retrofits for both historic and nonhistoric adobe buildings.

References

Bariola Bernales, Juan J.

1986 Dynamic stability of adobe walls. Ph.D. diss., University of Illinois at Urbana-Champaign.

Gavrilovic, P., V. Sendova, L. Taskov, L. Krstevska, E. Leroy Tolles, and W. S. Ginell

1996 Shaking table tests of adobe structures. Report, IZIIS 96-36, Institute of Earthquake Engineering and Engineering Seismology, Skopje, Republic of Macedonia, and Getty Conservation Institute, Los Angeles, June.

Meli, Roberto, Oscar Hernandez, and M. Padilla

1980 Strengthening of adobe houses for seismic actions. In *Proceedings of the Seventh World Conference on Earthquake Engineering.* Istanbul: Turkish National Committee on Earthquake Engineering.

Molino de Garcia, Maria Elena

1990 Prevención sísmica en las construcciones de adobe, en la cuidad de Guatemala después de los terremotos de 1917–1918. Paper presented at the Sixth International Conference on the Conservation of Earthen Architecture, Las Cruces, N.Mex., October 14–19.

Moncarz, Piotr D., and Helmut Krawinkler

1981 Theory and application of experimental model analysis in earthquake engineering. Report no. 50, John A. Blume Earthquake Engineering Center, Department of Civil Engineering, Stanford University, Palo Alto, Calif., June.

Scawthorn, Charles, and Ann M. Becker

1986 Relative benefits of alternative strengthening methods for low strength masonry buildings. In *Proceedings of the Third National Conference on Earthquake Engineering, Charleston, South Carolina.* Oakland, Calif.: Earthquake Engineering Institute.

State Historic Building Code

1990 State of California, Title 24, Building Standards, part 8, June.

Thiel, Charles C., Jr., E. Leroy Tolles, Edna E. Kimbro, Frederick A. Webster, and William S. Ginell

1991 GSAP—Getty Conservation Institute guidelines for seismic strengthening of adobe project: Report of first year activities. Report, Getty Conservation Institute, Marina del Rey, Calif., October 1.

Tolles, E. Leroy, and William S. Ginell

2000 Getty seismic adobe project (GSAP): Analysis of the results of shaking table tests on large scale model adobe structures. Report, Getty Conservation Institute, Los Angeles.

Tolles, E. Leroy, Edna E. Kimbro, and William S. Ginell

n.d. *Planning and Engineering Guidelines for the Seismic Stabilization of Historic Adobe Structures.* GCI Scientific Program Reports. Los Angeles: Getty Conservation Institute. Forthcoming.

Tolles, E. Leroy, Edna E. Kimbro, Charles C. Thiel Jr., Frederick A. Webster, and William S. Ginell
1993 GSAP—Getty Conservation Institute guidelines for seismic strengthening of adobe project: Report of second year activities. Report, Getty Conservation Institute, Marina del Rey, Calif., October 31.

Tolles, E. Leroy, and Helmut Krawinkler
1989 Seismic studies on small-scale models of adobe houses. Ph.D. diss., John A. Blume Earthquake Engineering Center, Department of Civil Engineering, Stanford University, Palo Alto, Calif.

Tolles, E. Leroy, Frederick A. Webster, Anthony Crosby, and Edna E. Kimbro
1996 *Survey of Damage to Historic Adobe Buildings after the January 1994 Northridge Earthquake*. GCI Scientific Program Reports. Los Angeles: Getty Conservation Institute.

Vargas-Neumann, Julio, Juan J. Bariola, and Marcial Blondet
1984 *Resistencia de la Mamposteria de Adobe*. Publication DI-84-01. Lima: Department of Engineering, Catholic University of Peru.

Vargas-Neumann, Julio, and G. Otazzi
1981 Investigaciones en adobe. In *Proceedings of the International Workshop on Earthen Buildings in Seismic Areas, Albuquerque, New Mexico*. Albuquerque: University of New Mexico.

Wiegel, Robert L., ed.
1970 *Earthquake Engineering*. Englewood Cliffs, N.J.: Prentice-Hall.

Glossary

adobe An outdoor, air-dried, unburned brick made from a clayey soil (10–30% clay content) and often mixed with straw or animal manure.

basal erosion A coving-type of deterioration at the base of an adobe wall.

bond beam A wood or concrete beam typically added to a wall at the roof level around the perimeter of a building.

convento Priest's residence (Spanish).

corredor Covered (roofed) exterior corridor or arcade, called a portal or portico in New Mexico, also referred to as a veranda or porch (Spanish).

cracked wall section A section of an adobe wall that is defined by a boundary of through-wall cracks.

diaphragm A large, thin structural element, usually horizontal, that is structurally loaded in its plane. It is usually an assemblage of elements that can include roof or floor sheathing, framing members to support the sheathing, and boundary or perimeter members.

EPGA Estimated peak ground acceleration.

epicenter The point on the ground surface directly above the hypocenter of an earthquake.

flexure Bending.

flexural stresses Stresses in an object that result from bending.

foundation settlement Downward movement of a foundation caused by subsidence or consolidation of the supporting ground.

freestanding walls Walls, such as garden walls, that are supported only laterally at the ground level. They have no roof or floor framing attached.

g Acceleration due to gravity at sea level (1 g = 32.2 ft/sec^2 or 980 cm sec^{-2}).

ground motion Lateral or vertical movement of the ground, as occurs in earthquakes.

HABS Historic American Building Survey. An ongoing federal documentation program of historic buildings in the United States, initiated as part of the Works Progress Administration (WPA) in the 1930s.

headers Adobe blocks placed with the long direction perpendicular to the plane of the wall.

hypocenter The "exact" point at which the ground begins to rupture at the beginning of an earthquake.

in plane Deflections or forces that are in the plane of a wall.

joists Closely spaced horizontal beams (spaced at approximately 2 feet [0.6 m] on center) that span an area, such as a floor or ceiling.

lintel A horizontal support beam over a window or door opening in a wall, usually made of wood in adobe buildings.

load-bearing Building elements, such as walls, that carry vertical loads from floors or roofs.

MMI Modified Mercalli Intensity scale. A qualitative measure of the local damage caused by an earthquake. Expressed in Roman numerals I through XII.

non-load-bearing Building elements, such as walls, that do not carry vertical loads from floors or roofs.

out of plane Deflections or forces that are perpendicular to the plane of a wall.

overturning Collapse of a wall caused by rotation of the wall about its base.

peak ground acceleration (PGA) The maximum ground acceleration produced by an earthquake. Often used to quantify the relative severity of an earthquake. Expressed in terms of g.

rafters Timbers or beams giving form, slope, and support to a roof.

Richter scale A measure of the total energy released by an earthquake. Expressed on a logarithmic scale.

shear forces Forces that can develop both in plane and out of plane and are caused by an opposite but parallel sliding motion of a body's planes. In adobe walls, such forces typically occur in the plane of the wall and cause diagonal cracking.

slenderness ratio (S_L) The ratio of the height of a wall to its thickness. Many historic adobe buildings have relatively thick walls ($S_L = 5$). Thin adobe walls ($S_L = 10$) are more susceptible to out-of-plane damage or collapse.

slumping Bulging at the base of an adobe wall resulting from loss of strength due to increased plasticity caused by moisture intrusion.

stretchers Adobe blocks placed with the long direction parallel to the plane of the wall.

tapanco Attic, loft, garret, or half-story of a building that is accessed by stairs or a ladder in the gable-end wall (Spanish).

wet/dry cycles Repeated cycles of water saturation and drying out that can lead to a loss of cohesion of the clay particles in adobe and that results in a weakened material.

wythe Portion of a wall that is one masonry unit thick.

Bibliography of Publications Resulting from GSAP

Gavrilovic, P., V. Sendova, L. Taskov, L. Krstevska, E. Leroy Tolles, and W. S. Ginell. "Shaking Table Tests of Adobe Structures." Report, IZIIS 96-36, Institute of Earthquake Engineering and Engineering Seismology, Skopje, Republic of Macedonia, and Getty Conservation Institute, Los Angeles, June 1996.

Gavrilovic, P., E. Leroy Tolles, and William S. Ginell. "Behavior of Adobe Structures during Shaking Table Tests and Earthquakes." In *Earthquake Engineering: Proceedings of the 11th European Conference, Paris, France, 6–11 September,* edited by P. Bisch, P. Labbé, and A. Pecker, 172. Rotterdam: A. A. Balkema, 1998.

Ginell, William S. "Seismic Stabilization of Historic Structures." *Conservation: The GCI Newsletter* 10, no. 3 (1995): 22.

———. "When the Earth Moves." *Conservation: The GCI Newsletter* 11, no. 1 (1996): 12–13.

Ginell, William S., Charles C. Thiel Jr., E. Leroy Tolles, and Frederick A. Webster. "Seismic Stabilization of Historic Adobe Buildings." In *Structural Studies of Historic Buildings 4.* Vol. 2, *Dynamics, Repairs, and Restoration,* 53–60. Southampton, U.K.: Computational Mechanics Publications, 1995.

Ginell, Willam S., and E. Leroy Tolles. "Preserving Safety and History: The Getty Seismic Adobe Project at Work." *Conservation: The GCI Newsletter* 14, no. 1 (1999): 12–14.

———. "Seismic Stabilization of Historic Adobe Buildings." Paper presented at the American Institute for Conservation Conference, Arlington, Va., June 1998. *Journal of the American Institute for Conservation,* forthcoming.

Kimbro, Edna E. "Conservation Principles Applied to Seismic Retrofitting of Culturally Significant Adobe Buildings." In *The Seismic Retrofit of Historic Buildings Conference Workbook,* edited by David W. Look, 6.1–12. San Francisco: Western Chapter of the Association for Preservation Technology, 1991.

———. "Conservation Principles Applied to Seismic Retrofitting of Culturally Significant Adobe Buildings." In *Seventh International Conference on the Study and Conservation of Earthen Architecture, Silves, Portugal, October 24 to 30, 1993,* edited by Margarida Alcada, 526–32. Lisbon: Direccão Geral dos Edificios e Monumentos Nacionais, 1993.

Levin, Jeffrey. "Adobes in the Seismic Zone." *J. Paul Getty Trust Bulletin* 8, no. 3 (1994): 6–7.

Thiel, Charles C. Jr., E. Leroy Tolles, Edna E. Kimbro, Frederick A. Webster, and William S. Ginell. "GSAP—Getty Conservation Institute Guidelines for Seismic Strengthening of Adobe Project: Report of First Year Activities." Report, Getty Conservation Institute, Marina del Rey, Calif., October 1, 1991.

Tolles, E. Leroy. "Seismic Retrofitting of Historic Adobes." In *Seventh International Conference on the Study and Conservation of Earthen Architecture, Silves, Portugal, October 24 to 30, 1993,* edited by Margarida Alcada, 533–38. Lisbon: Direccão Geral dos Edificios e Monumentos Nacionais, 1993.

Tolles, E. Leroy, and William S. Ginell. "Overview of the Getty Seismic Adobe Project." In *Workshop on the Seismic Retrofit of Historic Adobe Buildings, Held at the J. Paul Getty Museum, March 10, 1995,* 67–79. Pasadena, Calif.: Earthen Building Technologies, 1995.

———. "Getty Seismic Adobe Project (GSAP): Analysis of the Results of Shaking Table Tests on Large Scale Model Adobe Structures." Report, Getty Conservation Institute, Los Angeles, 2000.

Tolles, E. Leroy, Edna E. Kimbro, and William S. Ginell. *Planning and Engineering Guidelines for the Seismic Stabilization of Historic Adobe Structures.* GCI Scientific Program Reports. Los Angeles: Getty Conservation Institute, forthcoming.

Tolles, E. Leroy, Edna E. Kimbro, Charles C. Thiel Jr., Frederick A. Webster, and William S. Ginell. "GSAP—Getty Conservation Institute Guidelines for Seismic Strengthening of Adobe Project: Report of Second Year Activities." Report, Getty Conservation Institute, Marina del Rey, Calif., October 31, 1993.

Tolles, E. Leroy, Edna E. Kimbro, Frederick A. Webster, Anthony Crosby, and William S. Ginell. "Preliminary Survey of Historic Adobe Buildings Affected by the Northridge Earthquake of January 17, 1994." Report, Getty Conservation Institute, Marina del Rey, Calif., March 3, 1994.

Tolles, E. Leroy, Charles C. Thiel Jr., Frederick A. Webster, and William S. Ginell. "Advances in the Seismic Retrofitting of Adobe Buildings." In *Proceedings of the 5th U.S. National Conference on Earthquake Engineering: Papers Presented at the July 10–14, 1994, Conference in Chicago,* 277–82. Oakland, Calif.: Earthquake Engineering Research Institute, 1994.

Tolles, E. Leroy, Charles C. Thiel Jr., Frederick A. Webster, Edna E. Kimbro, and William S. Ginell. "Recent Developments in Understanding the Seismic Performance of Historic Adobe Buildings." In *The Seismic Retrofit of Historic Buildings Conference Workbook,* edited by David W. Look, 7.1–32. San Francisco: Western Chapter of the Association for Preservation Technology, 1991.

Tolles, E. Leroy, Frederick A. Webster, Anthony Crosby, and Edna E. Kimbro. *Survey of Damage to Historic Adobe Buildings after the January 1994 Northridge Earthquake.* GCI Scientific Program Reports. Los Angeles: Getty Conservation Institute, 1996.

Webster, Frederick A. "Costa Rica Earthquakes, December 22, 1990, and April 22, 1991." Reconnaissance Report—Historic Structures, Getty Conservation Institute, Marina del Rey, Calif., July 1991.

About the Authors

E. Leroy Tolles has worked on the seismic design, testing, and retrofit of adobe buildings since the early 1980s, specializing in the structural design and construction of earthen and wood buildings. He received his doctorate from Stanford University in 1989, where his work focused on the seismic design and testing of adobe houses in developing countries. He has led multidisciplinary teams to review earthquake damage after the 1985 earthquake in Mexico, the 1989 Loma Prieta earthquake, and the 1994 Northridge earthquake. He was principal investigator for the Getty Conservation Institute's Getty Seismic Adobe Project and has coauthored numerous publications on seismic engineering. He is principal for ELT & Associates, an engineering and architecture firm in northern California.

Edna E. Kimbro is an architectural conservator and historian, specializing in research and preservation of Spanish and Mexican colonial architecture and material culture of early California. She studied architectural history at the University of California, Santa Cruz, and, in 1976, rehabilitated the last remaining adobe building of the Spanish Villa de Branciforte. Through the 1980s, she was involved in restoration of the Mission Santa Cruz for the California Department of Parks and Recreation. In 1989, she attended UNESCO's International Centre for the Study of the Restoration and the Preservation of Cultural Property in Rome to study seismic protection of historic adobe buildings. In 1990, she became preservation specialist for the Getty Seismic Adobe Project. She is coauthor of a number of publications, and works as a project coordinator and preservation consultant for several historic sites in California.

Frederick A. Webster is a civil/structural engineer, working in California, who specializes in design, repair, and retrofitting of historic buildings. Since the early 1980s, he has researched, tested, lectured on, and designed adobe and rammed-earth buildings. He has also designed seismic retrofits, upgrades, repairs, and rehabilitation for historic and older adobes throughout the state. As part of the Getty Conservation Institute's earthquake reconnaissance team, he performed post-earthquake damage surveys of twenty historic adobes following the 1994 Northridge earthquake, and of historic and cultural property following more recent earthquakes in Costa Rica. He is currently responsible for the structural design of several adobe buildings to be constructed as part of a large winery complex in Santa Barbara County.

William S. Ginell is a materials scientist with extensive experience in industry. In 1943, after graduating from the Polytechnic Institute of Brooklyn with a bachelor's degree in chemistry, he became part of the secret research team at Columbia University working to develop the atomic bomb. After the war, he went on to receive his Ph.D. in physical chemistry from the University of Wisconsin and spent nine years at the Brookhaven National Laboratory on Long Island, New York, followed by twenty-six years working for aerospace firms in California. In 1984, he joined the Getty Conservation Institute and helped to design the laboratories at the GCI's Marina del Rey facility. He is currently senior conservation research scientist at the GCI in Los Angeles and was project director of the Getty Seismic Adobe Project.